Computational Anatomical Animal Models

Methodological developments and research applications

IPEM–IOP Series in Physics and Engineering in Medicine and Biology

About the Series

Series in Physics and Engineering in Medicine and Biology will allow IPEM to enhance its mission to 'advance physics and engineering applied to medicine and biology for the public good.'

Focusing on key areas including, but not limited to:

- clinical engineering
- diagnostic radiology
- informatics and computing
- magnetic resonance imaging
- nuclear medicine
- physiological measurement
- radiation protection
- radiotherapy
- rehabilitation engineering
- ultrasound and non-ionising radiation

A number of IPEM-IOP titles are published as part of the EUTEMPE Network Series for Medical Physics Experts.

Computational Anatomical Animal Models

Methodological developments and research applications

Habib Zaidi
Geneva University Hospital, Switzerland

IOP Publishing, Bristol, UK

ISBN 978-0-7503-1344-5 (ebook)
ISBN 978-0-7503-1345-2 (print)
ISBN 978-0-7503-1346-9 (mobi)

DOI 10.1088/2053-2563/aae1b4

Version: 20181201

IOP Expanding Physics
ISSN 2053-2563 (online)
ISSN 2054-7315 (print)

British Library Cataloguing-in-Publication Data: A catalogue record for this book is available from the British Library.

Published by IOP Publishing, wholly owned by The Institute of Physics, London

IOP Publishing, Temple Circus, Temple Way, Bristol, BS1 6HG, UK

US Office: IOP Publishing, Inc., 190 North Independence Mall West, Suite 601, Philadelphia, PA 19106, USA

Contents

Editor biography

Habib Zaidi

Professor Habib Zaidi, B.Eng, M.Sc, Ph.D, PD, FIEEE

Professor Habib Zaidi is Chief physicist and head of the PET Instrumentation & Neuroimaging Laboratory at Geneva University Hospital and faculty member at the medical school of Geneva University. He is also a Professor of Medical Physics at the University of Groningen (Netherlands), Adjunct Professor of Medical Physics and Molecular Imaging at the University of Southern Denmark, and visiting Professor at IAS/University Cergy-Pontoise (France). He was guest editor for 10 special issues of peer-reviewed journals and serves on the editorial board of leading journals in medical physics and medical imaging. He has been elevated to the grade of IEEE fellow and was elected liaison representative of the *International Organization for Medical Physics (IOMP)* to the World Health Organization (WHO). His academic accomplishments in the area of quantitative PET imaging have been well recognized by his peers and by the medical imaging community at large since he is a recipient of many awards and distinctions among which the prestigious *2003 Young Investigator Medical Imaging Science Award* given by the *Nuclear Medical and Imaging Sciences Technical Committee of the IEEE*, the *2004 Mark Tetalman Memorial Award* given by the *Society of Nuclear Medicine*, the *2007 Young Scientist Prize in Biological Physics* given by the *International Union of Pure and Applied Physics (IUPAP)*, the prestigious ($100 000) *2010 kuwait Prize of Applied sciences* (known as the *Middle Eastern Nobel Prize*) given by the *Kuwait Foundation for the Advancement of Sciences (KFAS)*, the 2013 *John S Laughlin Young Scientist Award* given by the *American Association of Physicists in Medicine (AAPM)*, the 2013 *Vikram Sarabhai Oration Award* given by the *Society of Nuclear Medicine, India (SNMI)*, the 2015 *Sir Godfrey Hounsfield Award* given by the *British Institute of Radiology (BIR)* and the *2017 IBA-Europhysics Prize* given by the European Physical Society (EPS). Professor Zaidi has been an invited speaker of over 150 keynote lectures and talks at an international level, has authored over 250 peer-reviewed articles in prominent journals and is the Editor of four textbooks including this volume. Email: habib.zaidi@hcuge.ch; Web: http://pinlab.hcuge.ch/.

List of contributors

Tianwu Xie
Geneva University Hospital, Switzerland

Habib Zaidi
Geneva University Hospital, Switzerland

Paul Segars
Duke University, USA

Akram Mohammadi
National Institute of Radiological Sciences (NIRS), National Institutes for Quantum and Radiological Science and Technology, Japan

Mitra Safavi-Naeini
Australian Nuclear Science and Technology Organisation (ANSTO), Australia

Sakae Kinase
Japan Atomic Energy Agency (JAEA)/Ibaraki University, Japan

Qian Liu
Huazhong University of Science and Technology, China

Guozhi Zhang
University Hospitals Leuven, Belgium

José-María Gómez-Ros
Research Centre for Energy, Environment and Technology, Spain

Choonsik Lee
National Institutes of Health, USA

Kanako Wake
National Institute of Information and Communications Technology, Japan

Akimasa Hirata
Nagoya Institute of Technology, Japan

Kenji Taguchi
Kitami Institute of Technology, Japan

Pedro Arce
Centro de Investigaciones Energéticas, Medioambientales y Tecnologicas (CIEMAT), Spain

Juan Ignacio Lagares
Medical Applications Unit, CIEMAT, Spain

Josep Sempau
Department of Physics, Universitat Politecnica de Catalunya, Spain

George Kagadis
University of Patras, Greece

Panagiotis Papadimitroulas
BioEmission Technology Solutions R&D Department, Greece

Niels Kuster
IT'IS Foundation and Swiss Federal Institute of Technology Zurich, Switzerland

Rameshwar Prasad
Rush University Medical Center, USA

M'hamed Bentourkia
Université de Sherbrooke, Canada

Mahdjoub Hamdi
University of Mostaganem, Algeria

Faiçal A Slimani
Université de Sherbrooke, Canada

Part I

Computational models

Chapter 1

Historical development and overview of computational animal models

Tianwu Xie and Habib Zaidi

1.1 Introduction

Laboratory animals have been widely used in preclinical research for the development and testing of new treatment strategies and imaging techniques and to investigate the exposure–response relationship between biological effectiveness and different types of non-ionizing and ionizing radiation. In this context, computational anthropomorphic animal models have been used as surrogates to characterize the physics and anatomy of laboratory animals to simulate radiation transport (such as microwaves, photons, electrons, and visible light, etc) and mimic the behavior of molecules/radiopharmaceuticals and medical imaging devices including positron emission tomography (PET), single-photon emission computed tomography (SPECT), bioluminescence tomography (BLT), fluorescence molecular tomography (FMT), micro-computed tomography (micro-CT), magnetic resonance imaging (MRI) and optical imaging (OI) [1, 2]. The simulated results based on computational animal models could provide the radiation dose delivered to biological tissues from both non-ionizing/and ionizing radiation exposures and generate image datasets for testing new image reconstruction and analysis algorithms for preclinical imaging platforms and evaluate the performance of medical imaging equipment before the construction of costly prototypes. In all applications, the foundation of performing animal-based preclinical simulation research is accurate anatomical and physiological modeling of laboratory animals [3, 4].

Overall, computational models can be divided into three types: stylized models defined by mathematical equations, voxel-based models constructed using digital volume arrays and boundary representation (BREP) models with boundary representations. The internal organs and exterior contours of stylized models are represented by simplified mathematical geometries, such as planes, spheres, ellipses, cones, cylinders and toroidal structures. Voxel-based models consist of matrices of

doi:10.1088/2053-2563/aae1b4ch1

segmented voxels representing different tissues and are commonly produced from tomographic images. BREP models employ surface representation methods [4] (such as polygon mesh or non-uniform rational B-spline (NURBS) surfaces) to describe the organs and body contours. The advantage of stylized models is the easy implementation and fast calculation within simulation codes. Their drawback is that they fail to reproduce accurately complex anatomic details. Voxel-based models provide better anatomic authenticity than stylized models but introduce inherent discretization errors by using voxel matrices. These models are also difficult to modify to describe new organ's contours. BREP models have the capability to deform and preserve most anatomic realism but cannot be directly used in most Monte Carlo simulation tools because of the complexity of surface representations [4].

By 2017, over 140 computational models had been reported in about 70 scientific articles for 15 different animals, including mice, rats, trout, crabs, frogs, flatfish, bees, deer, canines, earthworms, goats, ducks, monkeys, pigs and rabbits. Table 1.1 summarizes the computational animal models reported since 1994. In this table, anatomical features describe the characteristics of the animal specimen, model number refers to the number of reported unique animal models in the reference and imaging modality is the imaging method used to derive the animal images. Rat models, mouse models and models of other animals represent 46%, 38% and 16% of the total amount of computational animal models, respectively. With the exponentially increasing number of animal models and widespread recognition of computational modeling and simulation as trustworthy preclinical research tools, a number of questions are being addressed including how close the computational models are to the anatomy and physiology of real animals, how the simulated results change across different models, and what will be the future directions for computational modeling and simulation in preclinical research? Understanding of the processes of computational modeling and simulations for a variety of applications will help us to answer these questions. In this chapter, we will summarize the history of computational animal models as well as their development during the last 20 years and give some insight into future directions for computational modeling and simulations in future preclinical research.

1.2 Construction of computational models

According to the simulation tasks, requirements in terms of computational efficiency and the geometrical compatibility of simulation tools, various solid-geometry shapes can be employed in computational modeling to represent detailed anatomical features of laboratory animals and assigned multiple physicochemical characteristics, such as chemical composition, tissue density, relative permittivity, electric conductivity, or scattering coefficients for different radiation types.

As an example, the anatomical measurements of the left lung and right lung are obtained from medical images of rats and used to determine the semi-principal axes, the semi-major axis and semi-minor axis of the appropriate ellipses to depict the lungs of stylized rat models. Figure 1.1(a) shows the lung model of the rat [5] described by the union of two ellipsoids. Figure 1.1(b) shows the stylized lung model after subtracting

Table 1.1. Summary of reported computational models for laboratory animals.

Year	Developer	Affiliation of authors	Anatomical features	Model number	Imaging modality	Model type
1994	Hui *et al* [6]	Pacific Northwest Laboratory, USA	25 g nude mouse	1	Anatomic data	Stylized model
1999	Kennel *et al* [7]	Oak Ridge National Laboratory, USA	30 g mouse	1	Anatomic data	Stylized model
2001	Dhenain *et al* [8]	California Institute of Technology, USA	Mouse embryo	1	MRI	Voxel model
2001	Flynn *et al* [9]	Royal Free and University College Medical School, UK	Mouse	1	Anatomic data	Stylized model
2001	Wang *et al* [10]	Nagoya Institute of Technology, Japan	Mouse	1	MRI	Voxel model
2003	Kolbert *et al* [11]	Memorial Sloan-Kettering Cancer Center, USA	25 g female athymic mouse	1	MRI	Voxel model
2004	Funk *et al* [12]	University of California (San Francisco), USA	20 g, 30 g and 40 g mice with fixed axis ratios	3	Anatomic data	Stylized model
2004	Hindorf *et al* [13]	Lund University, Sweden	24 g mouse	1	Anatomic data	Stylized model
2004	Segars *et al* [14]	Johns Hopkins University, USA	Male C57BL/6 mouse	1	MRI	Hybrid model
2005	Miller *et al* [15]	University of Missouri, MO	25 g nude mouse with improved bone marrow model	1	Anatomic data	Stylized model
2006	Sato *et al* [16]	Chiyoda Technol Corporation, Japan	Mouse	1	Anatomic data	Stylized model
2006	Stabin *et al* [17]	Vanderbilt University, USA	27 g transgenic mouse	1	CT	Voxel model
2007	Bitar *et al* [18]	INSERM, France	30 g female athymic nude mouse	1	Cryosection	Voxel model
2007	Dogdas *et al* [19]	University of Southern California, USA	28 g normal male nude mouse	1	PET, CT and cryosection	Voxel model
2009	Zhang *et al* [20]	Chinese Academy of Sciences, China	20 g male BALB/c mouse	1	CT	Voxel model

(Continued)

2010	Keenan *et al* [21]	Vanderbilt University, Tennessee	25 g, 30 g and 35 g male C57BL/6 mice	3	Derived from MOBY (MRI)	Voxel model
2010	McIntosh *et al* [22]	Australian Centre for Radiofrequency Bioeffects Research, Australia	30 g ICR and ddY male, 22 g female mice, 22 g pregnant mouse and 0.5 g mouse fetus	4	Anatomic data	Voxel model
2010	Chow *et al* [23]	University of Toronto, Canada	Mouse	3	CT	Voxel model
2011	Larsson *et al* [24]	Lund University, Sweden	22 g, 28 g and 34 g male C57BL/6 mice with tumor model	3	Derived from MOBY (MRI)	Voxel model
2011	Mohammadi *et al* [25]	JAEA, Japan	Male C57BL/6 mouse	1	Derived from MOBY (MRI)	Voxel model
2011	Zhang *et al* [26]	Beijing Institute of Radiation Medicine, China	28 g mouse	1	Cryosection	Voxel model
2013	Xie and Zaidi [27]	Geneva University Hospital, Switzerland	Male C57BL/6 mouse with various body sizes	17	Derived from MOBY (MRI)	Voxel model
2013	Mauxion *et al* [28]	Centre de Recherche en Cancérologie de Toulouse, France	Male C57BL/6 mouse	1	Derived from MOBY (MRI)	Voxel model
2015	Welch *et al* [29]	Columbia University, NY	28 g nude male mouse	2	Derived from Digimouse atlas (CT and cryosection)	Voxel model
2017	Locatelli *et al* [30]	Université Paris Descartes-Sorbonne Paris Cité, France	21 g male C57BL/6 mouse and 20 g female C57BL/6 mouse	2	Derived from MOBY (MRI)	Voxel model
1997	Burkhardt *et al* [31]	Swiss Federal Institute of Technology, Switzerland	Rat	1	MRI	Voxel model
1997	Lapin *et al* [32]	Northwestern University, USA	350 g SD rat	1	CT	Voxel model
1999	Mason *et al* [33]	Systems Research Laboratories, USA; Brooks Air Force Base, USA	370 g rat	1	MRI	Voxel model
1999	Chou *et al* [34]	City of Hope National Medical Center, USA	SD rat	1	CT	Stylized model

2000	Watanabe *et al* [35]	Ministry of Posts and Telecommunications, Japan	126 g, 263 g and 359 g Fischer 344 rat	3	CT	Voxel model
2004	Funk *et al* [12]	University of California (San Francisco), USA	200 g, 300 g and 400 g rats with fixed axis ratios	3	Anatomic data	Stylized model
2004	Konijnenberg *et al* [36]	Tyco Healthcare, Netherlands	386 g Wistar rat	1	Anatomic data	Stylized model
2004	Leveque *et al* [37]	Centre National de la Recherche Scientifique, France	Rat	1	CT/MRI	Voxel model
2004	Schonborn *et al* [38]	Swiss Federal Institute of Technology, Switzerland	300 g male Wistar rat and 370 g male SD rat,	3	CT/MRI	Voxel model
2004	Segars *et al* [14]	Johns Hopkins University, USA	Male Wistar rat	1	MRI	Hybrid model
2006	Stabin *et al* [17]	Vanderbilt University, USA	248 g SD rat	1	CT	Voxel model
2006	Kainz *et al* [39]	Center for Devices and Radiological Health (CDRH), USA	567 g, 479 g, 252 g and 228 g SD rat	4	MRI, cryosection	Voxel model
2006	Lopresto *et al* [40]	ENEA, Italy	Rat	1	Anatomic data	Stylized model
2008	ICRP [41]	ICRP	314 g rat	1	Anatomic data	Stylized model
2008	Peixoto *et al* [42]	Universidade Federal de Pernambuco, Brazil	310 g Wistar rat	1	CT	Voxel model
2008	Wu *et al* [43]	HUST, China	156 g SD rat	1	Cryosection	Voxel model
2008	Wang *et al* [44]	Nagoya Institute of Technology, Japan	Pregnant Fischer 344 rat	1	MRI	Voxel model
2008	Pain *et al* [45]	Universités Paris 11/Paris 7, France	284 g SD rat	1	MRI	Voxel model
2009	Zhang *et al* [46]	HUST, China.	156 g Sprague–Dawley rat	1	Cryosection	Hybrid model
2010	Keenan *et al* [21]	Vanderbilt University, Tennessee	200 g, 300 g, 400 g, 500 g and 600 g male Wistar rat	5	Derived from MOBY (MRI)	Voxel model
2010	Xie *et al* [5]	HUST, China	156 g SD rat	1	Cryosection	Stylized model
2011	Arima *et al* [47]	NICT, Japan	115 g, 314 g and 472 g rat	3	CT	Voxel model

(*Continued*)

2012	Larsson *et al* [48]	Lund University, Sweden	225 g, 250 g, 275 g and 300 g Brown Norway rat	4	Derived from MOBY (MRI)	Voxel model
2012	Xie *et al* [49]	HUST, China	153 g SD rat	1	cryosection	Voxel model
2013	Xie and Zaidi [50]	Geneva University Hospital, Switzerland	Wistar rat at different ages	10	Derived from MOBY (MRI)	Voxel model
2013	Xie and Zaidi [51]	Geneva University Hospital, Switzerland	Wistar rat with different degrees of emaciation and obesity	7	Derived from MOBY (MRI)	Voxel model
2017	Locatelli *et al* [30]	Université Paris Descartes-Sorbonne Paris Cité, France	35 g, 208 g and 537 g female rat and 38 g, 277 g and 530 g male C57BL/6 rat	4	Derived from MOBY (MRI)	Voxel model
2008	ICRP [41]	ICRP	754 g crab	1	Anatomic data	Stylized model
2013	Caffrey *et al* [52]	Oregon State University, USA	464 g Dungeness crab	1	CT and MRI	Voxel model
2008	ICRP [41]	ICRP	1260 g trout	1	Anatomic data	Stylized model
2014	Hess *et al* [53]	Oregon State University, USA	658 g rainbow trout	1	CT and MRI	Voxel model
2014	Martinez *et al* [54, 55]	Clemson University, USA	~200 g rainbow trout	3	CT	Stylized/ hybrid/ voxel model
2008	ICRP [41]	ICRP	31.4 g frog	1	Anatomic data	Stylized model
2012	Kinase *et al* [56]	JAEA, Japan	33.7 g frog	1	cryosection	Voxel model
2008	ICRP [41]	ICRP	1310 g flatfish	1	Anatomic data	Stylized model
2013	Caffrey *et al* [57]	Oregon State University, USA	1024 g Pleuronectid flatfish	1	CT and MRI	Voxel model
2008	Padilla *et al* [58]	University of Florida, Florida	24 kg hound cross	3	CT	Stylized/ hybrid/ voxel model
2012	Kramer *et al* [59]	National Internal Radiation Assessment section, Radiation Protection Bureau, Canada	14 kg and 34 kg Doberman dogs	2	CT	Voxel model
2015	Stabin *et al* [60]	Vanderbilt University, TN	9.5 kg male and 8.3 kg female beagle dogs	2	CT	Hybrid model
2008	ICRP [41]	ICRP	0.589 g bee	1	Anatomic data	Stylized model
2016	Gomez *et al* [61]	Oregon State University, USA	*Apis mellifera* honeybee	1	CT	Voxel model
2008	ICRP [41]	ICRP	245 kg deer	1	Anatomic data	Stylized model

2008	ICRP [41]	ICRP	1260 g duck	1	Anatomic data	Stylized model
2008	ICRP [41]	ICRP	5.24 g earthworm	1	Anatomic data	Stylized model
1999	Mason *et al* [33]	Systems Research Laboratories, USA; Brooks Air Force Base, USA	20 kg pigmy goat	1	MRI	Voxel model
1999	Mason *et al* [33]	Systems Research Laboratories, USA; Brooks Air Force Base, USA	7.1 kg rhesus monkey and a phantom monkey	2	MRI	Voxel model
2008	Toivonen *et al* [62]	Radiation and Nuclear Safety Authority, Finland	25 kg pig	1	CT and MRI	Voxel model
2007	Wake *et al* [63]	National Institute of Information and Communications Technology, Japan	Rabbit	1	CT	Voxel model
2015	Caffrey *et al* [64]	Oregon State University, USA	2 kg male and 0.8 kg juvenile female black-tailed jackrabbits (*Lepus californicus*)	2	CT	Voxel model

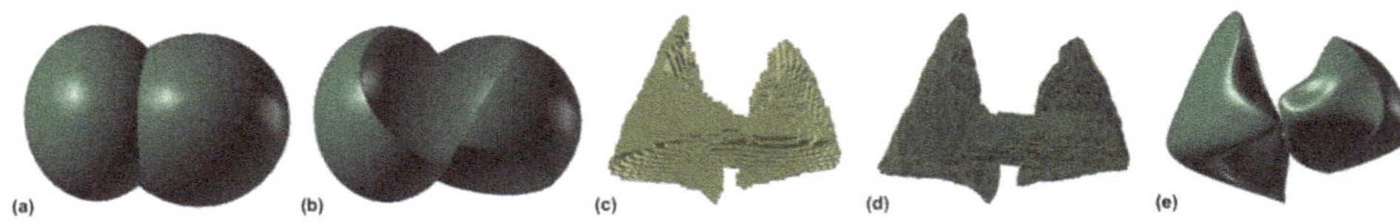

Figure 1.1. The rat lung model defined by different methods. (a) The stylized solid geometry before subtraction operation. (b) Lung model after the subtraction of the heart and liver. (c) The voxel-based model of the lung. (d) The polygon mesh model of the lung. (e) The NURBS model of the lung. Reproduced from [65]. John Wiley & Sons. © 2016 American Association of Physicists in Medicine.

the surrounding organs (such as the heart and liver) using Boolean operations. Using the same approach, a series of simple mathematical formulations can be conducted to represent the internal organs and body contours of animals. The approximate description of organs using mathematical equations is geometrically compatible and computationally efficient for most simulation tools but introduces significant anatomical uncertainties because of the considerably overlooked anatomical details.

The rectangular cuboids used in voxel models can be directly converted from a medical image dataset, which also facilitates the geometry representation in simulation tools. However, the construction of voxel models based on image data sets (such as high resolution cryosection images, CT, PET or MR images) requires dedicated segmentation to assign each rectangular cuboid an ID corresponding to a particular tissue of interest and can take a significant amount of time when manual segmentation is performed. Overall, manual segmentation can be applied to define all regions of interest in the body but the process is time-consuming and requires excellent knowledge of the anatomy, therefore, a number of automated or semi-automated segmentation approaches (e.g. thresholding, variational approaches, machine learning and techniques based on multi-atlas segmentations, etc) and image processing software (e.g. 3D-Doctor, Photoshop, ImageJ, etc) have been developed and used to alleviate the workload for tissue identification [66]. Automated segmentation techniques were deemed impractical for internal tissues having similar density but can perform well for segmenting the skeleton, body contour and cavity regions. As shown in figure 1.1(c), the ladder-shaped boundary of the voxel-based lung model introduces an approximate description of smooth organ surface where the anatomical accuracy of the model depends on the voxel resolution. In the meantime, the adjustment and deformation of organs and body contours in voxel models is difficult because of the rigid structure of the voxel array. The resolution of adopted medical image datasets determines the attainable model spatial resolution of generated voxel models. For voxel models based on micro-MR images, the spatial resolution is about 25 μm to 100 μm when high strength magnetic fields are used. For voxel models based on micro-CT and cryosection dataset, the spatial resolution can be about 50 μm and 20 μm, respectively.

Internal organs and body of organisms can also be described using advanced boundary representation techniques, such as polygon meshes and B-Splines. Figures 1.1(d) and (e) illustrate the mesh model and NURBS model of rat lung constructed based on the binary voxel model, by using the marching cubes algorithm and 3D modeling software, respectively. These BREP models can describe smooth organ's surface and provide better anatomical realism than both voxel model and stylized model. BREP models using surface representation, such as NURBS and polygon mesh to describe organisms are capable of holding the flexibility of deforming models to mimic reliable anatomic characteristics. The surface representation of BREP models can be modified to generate models of certain anatomical or physiological characteristics, such as obese, emaciated or tumor-bearing animals. Using the voxelizing process, the generated BREP models can be adopted within most simulation tools. The customization capability of BREP models greatly promoted the utilization of dedicated computational models in preclinical applications, imaging physics and radiation dosimetry research.

In summary, stylized models are based on mathematical equations, voxel models are based on segmentation of tomographic images while BREP models are based on advanced surface representations and are usually more sophisticated than stylized and voxel models. The stylized, voxel and hybrid models developed based on the same rat dataset are shown and compared in figure 1.2. As shown in figure 1.3, the total number of computational models for laboratory animals has increased by a factor of 3.3 since 1994 while voxel models, stylized models and BREP models represent 77%, 18% and 5% of all reported computational animal models, respectively.

1.3 Overview of existing computational animal models

Computational animal models evolved from tissue-equivalent homogeneous ellipsoids to realistic anthropomorphic models which mimic organisms in terms of composition, anatomy and physiology. An understanding of the history of computational models of small animals provides an insight into its future development. Overall, the reported computational models can be divided into three major categories based on the represented species: (i) mouse models; (ii) rat models; and (iii) models of other animals.

1.3.1 Mouse models

In preclinical research, the laboratory mouse is the mammalian model most commonly used to evaluate the pharmacokinetics, biodistribution, radiotoxicity

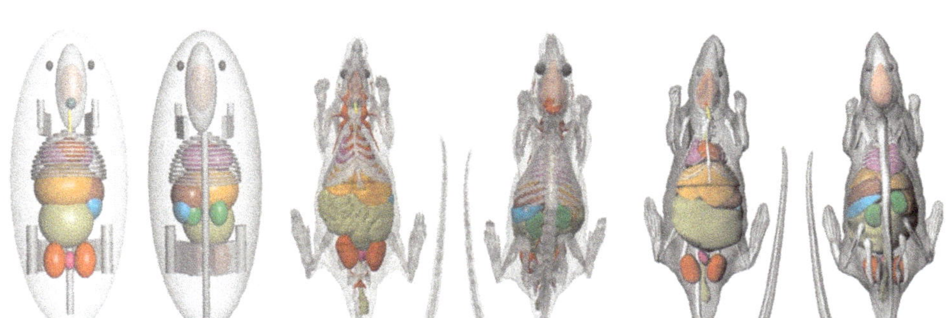

Figure 1.2. Dorsal–ventral views (right) and ventral–dorsal views (left) of (a) a stylized rat model, (b) a voxel model and (c) a hybrid model. Reproduced from [65]. John Wiley & Sons. © 2016 American Association of Physicists in Medicine.

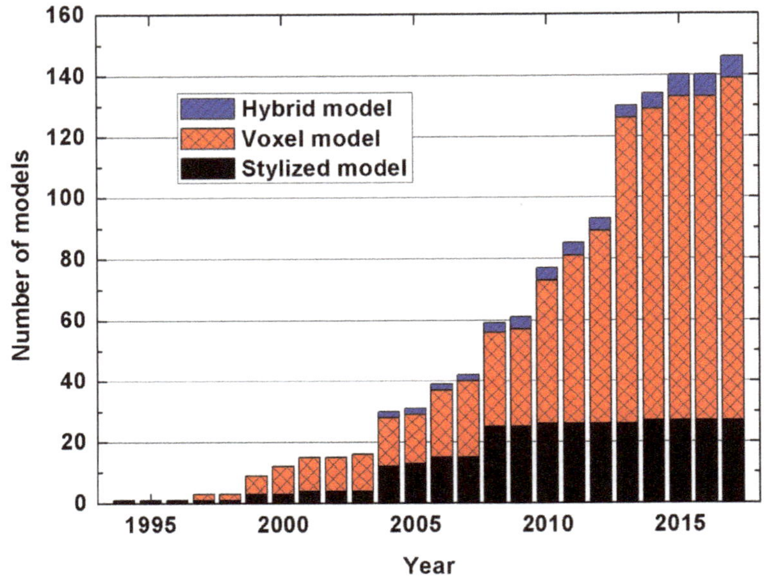

Figure 1.3. Number of developed computational animal models since 1994 reflecting the surveyed models summarized in table 1.1.

and effectiveness of new radiopharmaceuticals for molecular imaging, radiation therapy and many other scientific disciplines. In longitudinal studies, experimental animals are used in serial studies and the same animal may serve as its own control. Therefore, the delivered radiation dose from radiological preclinical imaging procedures (e.g. micro-SPECT/CT or micro-PET/CT) might be very high. Overall, the delivered radiation dose to small animals during hybrid imaging experiments is nonlethal but may induce changes in biological pathways which may affect the experimental outcomes [67]. Typical radiation doses to laboratory mouse from micro-CT imaging can range between 0.02 Gy and 1.5 Gy depending on the required image quality [68–71] and imaging protocol [72]. Conversely, the radiation dose to a mouse from radiopharmaceuticals is determined by the used isotope, administered activity and mouse weight [27]. Using ellipsoids, Hui *et al* [6] constructed the first computational small-animal model of a laboratory mouse in 1994 (stylized model as shown in figure 1.4) and calculated the internal absorbed doses of Y-90 labeled immunoconjugates. Henceforward, various computational mouse models have been reported in the literature. Figure 1.5 shows representative mouse models including the stylized model of Hindorf *et al* [13], the widely used voxel model (Digimouse [19]) and the first BREP mouse model (MOBY [14]). Other well-known stylized mouse models were also reported including a 30 g tumor-bearing model by Kennel *et al* [7], a model with advanced kidney structures and tumors developed by Flynn *et al* [9], mouse models with fixed axis ratios [12] and a mouse model with detailed bone marrow structures [15]. A number of voxel-based mouse models were also developed including a BALB/c mouse [20], an athymic nude mouse [11, 18, 19], a C57BL/6 mouse [21, 24, 25, 27], ddY mice and ICR mice [22] of both genders and different

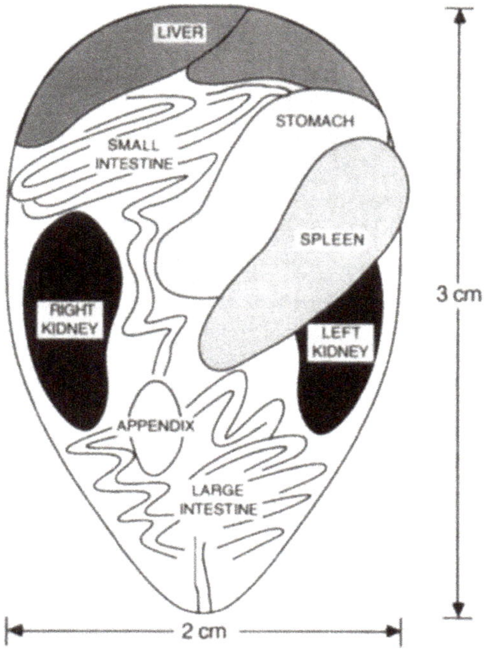

Figure 1.4. The computational model developed by Hui *et al* [6]. (Reproduced with permission from reference [6]. Copyright © 1994 American Cancer Society.)

Figure 1.5. (a) The stylized mouse model [13], (b) the Digimouse voxel model [19] and (c) the MOBY model [14]. (Reproduced with permission from references [13, 14, 19].)

weights where Dogdas *et al* [19] developed the well-known Digimouse model from cryosection, CT and PET images of a 28 g nude male mouse and McIntosh *et al* [22] developed a series of pregnant mouse models and a mouse fetus. In 2004, the first BREP model of a mouse (MOBY) was developed by Segars *et al* [14] based on MR images of a male mouse. This model was used to generate a series of voxel-based mouse models [21], such as a tumor-bearing mouse [24] and models of various body weights [25] and sizes [27]. More recently, Wang *et al* [73] reported a deformable mouse atlas with adaptable body poses and weight.

1.3.2 Rat models

The rat is prized as the pre-eminent laboratory animal in biomedical research because of the similarities between the rodent genome and the human genome, and the easy and relatively low cost breeding of laboratory rats. The Wistar rat, Sprague–Dawley (SD) rat and Fischer rat are the most popular strains in preclinical research. Figure 1.6 compares a representative stylized model [36], voxel-based model [42] and BREP model [46] of laboratory rat.

A number of simplified stylized rat models were constructed based on the SD rat for ionizing [5, 12, 36, 41] and non-ionizing [34, 40] radiation dosimetry. In 1997, Lapin *et al* [32] and Burkhardt *et al* [31] developed voxel-based rat models based on CT and MR images, respectively, for electromagnetic field (EMF) dosimetry. Other voxel-based rat models were also developed based on CT and MRI for dosimetric analysis in electromagnetic modeling experiments [33, 37, 38]. In general, CT images were the basis for constructing most voxel-based models, such as the reported 248 g SD rat [17], a 310 g Wistar rat [42], three Fischer 344 rats of different weights [35]. Some models were developed using MR images, including the voxel models SD rat reported by Pain *et al* [45] and the pregnant rat model [44]. Wu *et al* [43] developed a SD rat model from cryosection images, whereas Kainz *et al* [39] developed voxel rat models based on the microtome slice pictures. The first BREP rat model, known as ROBY, was developed by Segars *et al* [14] based on the Wistar rat. In various applications, the ROBY model was modified to generate new voxel models, such as five 200–600 g Wistar rat models [21] and the model of a Brown Norway rat [48]. Using the original ROBY model, Xie and Zaidi [50, 51] developed a series of voxel-based rat models at different levels of emaciation and obesity (figure 1.7) and models at different ages. Another BREP rat model was reported by Xie *et al* [5] and Zhang *et al* [46] from Huazhong University of Science and Technology (HUST) based on cryosection images of an SD rat.

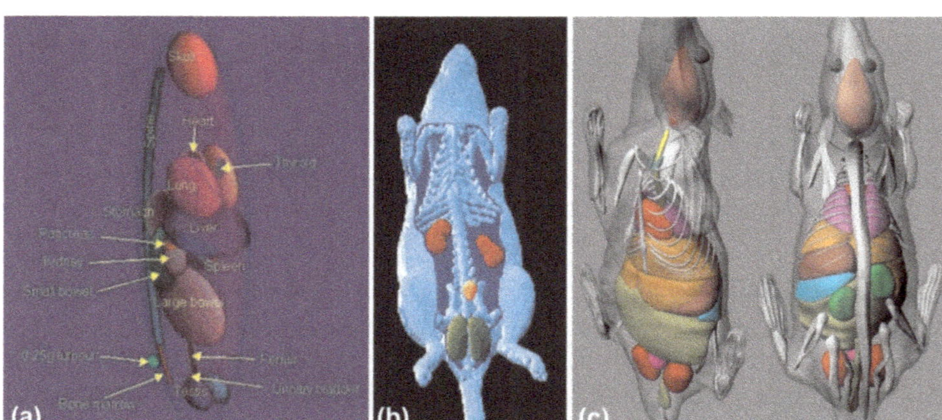

Figure 1.6. Comparison between three categories of computational models showing: (a) a stylized rat model [36], (b) a voxel-based rat model [42] and (c) the BREP rat model [46]. (Reproduced with permission from references [36, 42, 46].)

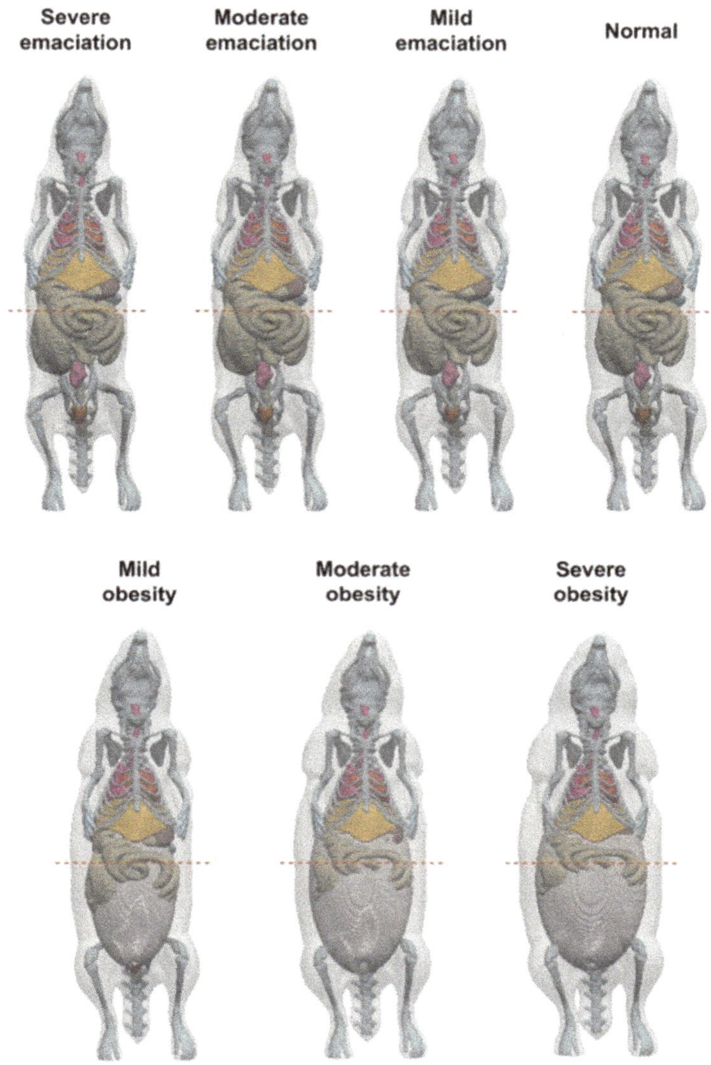

Figure 1.7. Computational models with different degrees of emaciation and obesity for rats. (Reproduced with permission from reference [51].)

1.3.3 Models of other animals

Except for mice and rats, few animal models of other smaller or larger organisms were developed for radiation protection and biomedical research. Figure 1.8 shows the deer model used by ICRP [41], the voxel trout model [53] and the BREP canine model [58]. The body of a deer is roughly represented by a 130 × 60 × 60 cm ellipsoid. In ICRP publication 108, eight stylized animal models: 0.589 g bee, 5.24 g earthworm, 31.4 g frog, 754 g crab, 1260 g trout, 1260 g duck, 1310 g flatfish and 245 kg deer were reported and used to study the radiation exposure of animals in the wild. Mason *et al* [33] reported a 20 kg pigmy goat model and two monkey models

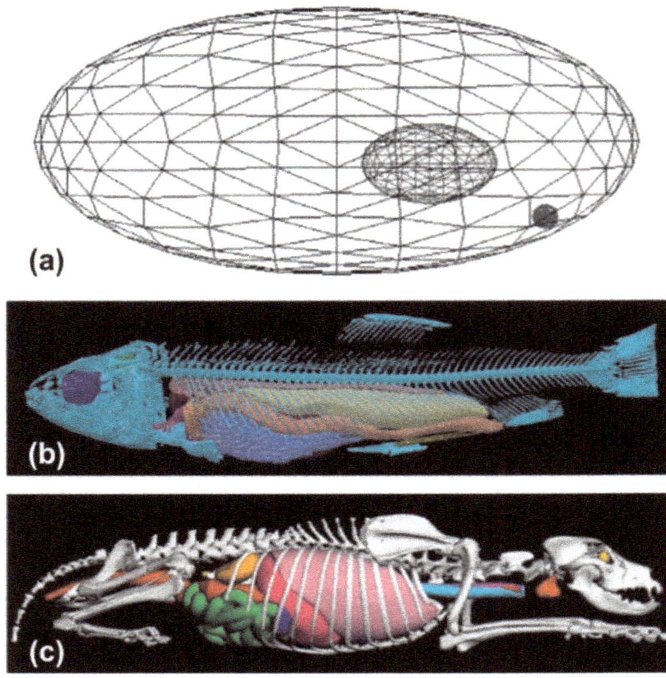

Figure 1.8. (a) The stylized deer model used by ICRP [41], (b) the voxel trout model [53], and (c) the BREP canine model [58]. (Reproduced with permission from references [41, 53, 58].)

based on MR images. Researchers from Oregon State University developed voxel models for 1024 g Pleuronectid flatfish, 658 g Rainbow trout and a 464 g Dungeness crab [53, 57, 74], using MR and CT images. A voxel-based frog model was also reported based on cryosection images [56]. For estimating absorbed dose from EMF radiations, Wake *et al* [63] and Toivonen *et al* [62] developed a rabbit model and a pig model in their research, respectively. Padilla *et al* [58] from the University of Florida developed a stylized model, a voxel-based model and a BREP model based on the same hound cross canine specimen.

1.4 Popular simulation tools for computational models

Computational phantoms are used to mimic the anatomic and physicochemical characteristics of laboratory animals and commonly integrated in simulation tools to perform radiation transport calculations inside the organisms, thus enabling the calculation of energy deposition and evaluation of the impact of physical degrading factors on the performance of medical imaging systems.

The Monte Carlo method is the most popular technique for ionizing radiation dosimetry in preclinical research where all aspects of particle interactions within heterogeneous biological materials [75, 76] can be accounted for based on statistical modeling. Popular Monte Carlo codes include the EGSnrc Monte Carlo system developed by Stanford Linear Accelerator Center for the purpose of simulating the transport of photons and electrons [77], the MCNP system developed by Los

Alamos National Laboratory with the capability of simulating 34 different types of particle [78] and the Geant4 software package developed by the European Organization for Nuclear Research [79]. In addition, many other simulation tools have been devised from Geant4, EGSnrc or MCNP for medical physics applications and light propagation simulations [80].

In electromagnetic field simulations, computational phantoms are commonly integrated within commercial or open-source software (such as XFDTD, CST, Meep ...etc) [81–83] to calculate the energy deposition in laboratory animals from non-ionizing radiation exposures. The finite-element methods (FEMs), boundary-element methods (BEMs) and the FDTD algorithm are popular numerical simulation methods for electromagnetism computations.

1.5 Summary

The historical development of computational animal models started from stylized models based on simplified mathematical formulas, through voxel-based models constructed from segmenting medical image datasets of real animals, and to BREP models which employ surface representations to hold the flexibility of deforming objects thus achieving anatomical realism of organs/bodies. The degree of complexity and realism of the reported computational animal models have been significantly improved. Currently, more than 140 computational models covering 15 different species of laboratory animals have been reported and used in preclinical imaging physics developments, non-ionizing and ionizing radiation dosimetry and almost all areas of medical physics. In all the reported models, more than 70% of them are voxel-based models while only seven hybrid animal models have been developed for mice, rats, trout and canines.

The advantages and disadvantages of the three types of computational models can be approximately elaborated under the themes of geometrical flexibility, anatomical fidelity and compatibility with the available simulation tools. Substantial deviations for dosimetric calculations have been reported in stylized models because of the unrealistic anatomy [5]. Voxel-based models and BREP models achieved much better anatomical realism than stylized models. However, the voxel resolution adopted in voxel models determines its anatomical fidelity and the cube-shaped organ surface may introduce voxel size-related uncertainty in dosimetry calculations [3]. Stylized and BREP models describe smooth organ surfaces and hold the flexibility of deforming to describe organ motions during heart pumping and respiratory movement.

Regarding compatibility of computational models with popular simulation tools, both voxel and stylized models can be easily integrated within commercial or open-source simulation platforms because of the adopted simple geometric elements. However, most BREP models, except tetrahedral-mesh-based models [84], cannot be directly imported into popular Monte Carlo tools. Two methods were used to improve the computational compatibility of BREP models. The first option is converting the BREP model to the corresponding voxel model and its use in simulation tools. The second solution involves upgrading the inherent geometry definition of simulation tools to handle the BREP models. Computer-aided design (CAD)-based polygon mesh models can be directly used in the new version of MCNP and Geant4 to implement general-purpose Monte Carlo simulations [85].

Another important issue for preclinical applications is the concept of using reference models, which have been introduced from the field of radiation protection for human beings. A series of reference animal models were reported by the ICRP and have been used for dose calculation in environmental protection. However, significant divergences of radiation dose have been reported between different animal models and animal subjects [27, 86]. In this context, for radiation dosimetry in preclinical research, the most appropriate and best-fitting computational animal model should be considered according to the geometrical characteristics of the adopted experimental specimen.

Compared to the real animals or physical phantoms, computational models enable users to define the desired physiological and anatomical properties in experimental subjects. The data from computational models were generated under controlled conditions and can be compared with the ground truth to assess the accuracy of image analysis and reconstruction methods and evaluate the performance of medical imaging devices. Another advantage of using computational models is that they can provide a large number of experimental subjects for medical physics studies, by modifying the anatomies and physiological characteristics of the model, thus reducing the cost of the experiment and the administrative complexity involved in clearing ethical aspects and solving logistic issues.

The selection of the appropriate model is the pivotal issue for the given applications involving the usage of computational models. For radiation dosimetry studies, the organ mass and the distance between the organs determine the calculated dosimetric results, thus, the animal model having the closest anatomy characteristics, including strain, length, age and body weight, as the experimental subject, would be the most appropriate one. In large-scale radiation dosimetry research (such as environmental radiation protection), the accuracy of dosimetric calculation is restricted by anatomic diversification between different subjects. Therefore, stylized models can be used in these applications to provide an average approximation for the investigated animal. The BREP model holds the flexibility of deforming objects to mimic cardiac and respiratory motions and would be suitable for advanced molecular imaging applications involving dynamic changes of anatomic structures within a living organism. Real laboratory animals are commonly used in basic and translational studies and preclinical research to investigate the biological response of radiopharmaceuticals and different radiation types.

All in all, with the rapid development of molecular multimodality imaging, more and more radiotracers are being developed and tested for imaging and targeted therapy application. Computational animal models will definitely play an important role in translational research for the validation of new radiopharmaceuticals and imaging techniques.

BREP models are capable of describing the realistic anatomy of living organisms and hold the flexibility enabling deforming objects for anatomic matching of experimental subjects. With the development of computer graphics tools and advanced simulation platforms, it is not far-fetched that we will witness an explosion in the development and application of hybrid animal models in multimodality imaging research and radiation dosimetry.

Acknowledgements

This work was supported by the Swiss National Science Foundation under grant SNSF 31003A-149957 and Geneva Cancer League.

References

[1] Xu X G 2014 An exponential growth of computational phantom research in radiation protection, imaging, and radiotherapy: A review of the fifty-year history *Phys. Med. Biol.* **59** R233–302

[2] Zaidi H and Xu X G 2007 Computational anthropomorphic models of the human anatomy: The path to realistic Monte Carlo modeling in radiological sciences *Annu. Rev. Biomed. Eng.* **9** 471–500

[3] Xu X G and Eckerman K F 2010 *Handbook of anatomical models for radiation dosimetry* (Boca Raton, FL: CRC Press)

[4] Zaidi H and Tsui B M 2009 Review of computational anthropomorphic anatomical and physiological models *Proc. IEEE* **97** 1938–53

[5] Xie T, Zhang G, Li Y and Liu Q 2010 Comparison of absorbed fractions of electrons and photons using three kinds of computational phantoms of rat *Appl. Phys. Lett.* **97** 033702

[6] Edmond Hui T, Fisher D R, Kuhn J A, Williams L E, Nourigat C and Badger C C *et al* 1994 A mouse model for calculating cross-organ beta doses from yttrium-90-labeled immunoconjugates *Cancer.* **73** 951–7

[7] Kennel S J, Stabin M, Yoriyaz H, Brechbiel M and Mirzadeh S 1999 Treatment of lung tumor colonies with ^{90}Y targeted to blood vessels: Comparison with the α-particle emitter ^{213}Bi *Nucl. Med. Biol.* **26** 149–57

[8] Dhenain M, Ruffins S W and Jacobs R E 2001 Three-dimensional digital mouse atlas using high-resolution MRI *Dev. Biol.* **232** 458–70

[9] Flynn A, Green A, Pedley R, Boxer G, Boden R and Begent R H J 2001 A mouse model for calculating the absorbed beta-particle dose from ^{131}I-and ^{90}Y-labeled immunoconjugates, including a method for dealing with heterogeneity in kidney and tumor *Radiat. Res.* **156** 28–35

[10] Jianqing W and Fujiwara O 2001 A novel setup for small animal exposure to near fields to test biological effects of cellular telephones *IEICE Trans. Commun.* **84** 3050–9

[11] Kolbert K S, Watson T, Matei C, Xu S, Koutcher J A and Sgouros G 2003 Murine S factors for liver, spleen, and kidney *J. Nucl. Med.* **44** 784–91

[12] Funk T, Sun M and Hasegawa B H 2004 Radiation dose estimate in small animal SPECT and PET *Med. Phys.* **31** 2680–6

[13] Hindorf C, Ljungberg M and Strand S-E 2004 Evaluation of parameters influencing S values in mouse dosimetry *J. Nucl. Med.* **45** 1960–5

[14] Segars W P, Tsui B M, Frey E C, Johnson G A and Berr S S 2004 Development of a 4-D digital mouse phantom for molecular imaging research *Mol. Imaging Biol.* **6** 149–59

[15] Miller W H, Hartmann-Siantar C, Fisher D, Descalle M-A, Daly T and Lehmann J *et al* 2005 Evaluation of beta-absorbed fractions in a mouse model for ^{90}Y, ^{188}Re, ^{166}Ho, ^{149}Pm, ^{64}Cu, and ^{177}Lu radionuclides *Cancer Biother. Radiopharm.* **20** 436–49

[16] Sato Y, Yamabayashi H and Nakamura T 2008 Internal dose distribution of ^{90}Y beta-ray source implanted in a small phantom simulating a mouse *Radioisotopes* **57** 385–91
[17] Stabin M G, Peterson T E, Holburn G E and Emmons M A 2006 Voxel-based mouse and rat models for internal dose calculations *J. Nucl. Med.* **47** 655–9
[18] Bitar A, Lisbona A, Thedrez P, Maurel C S, Le Forestier D and Barbet J *et al* 2007 A voxel-based mouse for internal dose calculations using Monte Carlo simulations (MCNP) *Phys. Med. Biol.* **52** 1013
[19] Dogdas B, Stout D, Chatziioannou A F and Leahy R M 2007 Digimouse: A 3D whole body mouse atlas from CT and cryosection data *Phys. Med. Biol.* **52** 577
[20] Zhang X, Tian J, Feng J, Zhu S and Yan G (ed) 2009 An anatomical mouse model for multimodal molecular imaging *Annual International Conference of the IEEE Engineering in Medicine and Biology Society* (Piscataway, NJ: IEEE)
[21] Keenan M A, Stabin M G, Segars W P and Fernald M J 2010 RADAR realistic animal model series for dose assessment *J. Nucl. Med.* **51** 471–6
[22] McIntosh R L, Deppeler L, Oliva M, Parente J, Tambuwala F and Turner S *et al* 2010 Comparison of radiofrequency exposure of a mouse dam and foetuses at 900 MHz *Phys. Med. Biol.* **55** N111
[23] Chow J C, Leung M K, Lindsay P E and Jaffray D A 2010 Dosimetric variation due to the photon beam energy in the small-animal irradiation: A Monte Carlo study *Med. Phys.* **37** 5322–9
[24] Larsson E, Ljungberg M, Strand S-E and Jönsson B-A 2011 Monte Carlo calculations of absorbed doses in tumours using a modified MOBY mouse phantom for pre-clinical dosimetry studies *Acta Oncol.* **50** 973–80
[25] Mohammadi A and Kinase S 2011 Influence of voxel size on specific absorbed fractions and S-values in a mouse voxel phantom *Radiat. Prot. Dosim.* **143** 258–63
[26] Zhang X, Xie X, Cheng J, Ning J, Yuan Y and Pan J *et al* 2012 Organ dose conversion coefficients based on a voxel mouse model and MCNP code for external photon irradiation *Radiat. Prot. Dosim.* **148** 9–19
[27] Xie T and Zaidi H 2013 Monte Carlo-based evaluation of S-values in mouse models for positron-emitting radionuclides *Phys. Med. Biol.* **58** 169–82
[28] Mauxion T, Barbet J, Suhard J, Pouget J P, Poirot M and Bardiès M 2013 Improved realism of hybrid mouse models may not be sufficient to generate reference dosimetric data *Med. Phys.* **40** 052501
[29] Welch D, Harken A, Randers-Pehrson G and Brenner D 2015 Construction of mouse phantoms from segmented CT scan data for radiation dosimetry studies *Phys. Med. Biol.* **60** 3589
[30] Locatelli M, Miloudi H, Autret G, Balvay D, Desbrée A and Blanchardon E *et al* 2017 RODES software for dose assessment of rats and mice contaminated with radionuclides *J. Radiol. Prot.* **37** 214
[31] Burkhardt M, Spinelli Y and Kuster N 1997 Exposure setup to test effects of wireless communications systems on the CNS *Health Phys.* **73** 770–8
[32] Lapin G and Allen C (ed) Requirements for accurate anatomical imaging of the rat for electromagnetic modeling *Proc. of the 19th Annual Int. Conf. of the IEEE Engineering in Medicine and Biology Society* 1997 (Piscataway, NJ: IEEE)
[33] Mason P, Ziriax J, Hurt W and D'Andrea J 1999 3-dimensional models for EMF dosimetry *Electricity and Magnetism in Biology and Medicine* (Berlin: Springer) pp 291–4

[34] Chou C, Chan K, McDougall J and Guy A 1999 Development of a rat head exposure system for simulating human exposure to RF fields from handheld wireless telephones *Bioelectromagnetics* **20** 75–92

[35] Watanabe S, Mukoyama A, Wake K, Yamanaka Y, Uno T and Taki M 2000 Microwave exposure setup for a long-term in vivo study *Proc. of the Int. Symp. on Antennas and Propagation Japan*

[36] Konijnenberg M W, Bijster M, Krenning E P and de Jong M 2004 A stylized computational model of the rat for organ dosimetry in support of preclinical evaluations of peptide receptor radionuclide therapy with ^{90}Y, ^{111}In, or ^{177}Lu *J. Nucl. Med.* **45** 1260–9

[37] Leveque P, Dale C, Veyret B and Wiart J 2004 Dosimetric analysis of a 900-MHz rat head exposure system *IEEE Trans. Microw. Theor. Tech.* **52** 2076–83

[38] Schönborn F, Poković K and Kuster N 2004 Dosimetric analysis of the carousel setup for the exposure of rats at 1.62 GHz *Bioelectromagnetics* **25** 16–26

[39] Kainz W, Nikoloski N, Oesch W, Berdiñas-Torres V, Fröhlich J and Neubauer G *et al* 2006 Development of novel whole-body exposure setups for rats providing high efficiency, National Toxicology Program (NTP) compatibility and well-characterized exposure *Phys. Med. Biol.* **51** 5211

[40] Lopresto V, Pinto R, De Vita A, Mancini S, Galloni P and Marino C *et al* 2007 Exposure setup to study potential adverse effects at GSM 1800 and UMTS frequencies on the auditory systems of rats *Radiat. Prot. Dosim.* **123** 473–82

[41] ICRP 2008 Publication 108: Environmental protection – the concept and use of reference animals and plants *Ann. ICRP* **38** 1–242

[42] Peixoto P, Vieira J, Yoriyaz H and Lima F 2008 Photon and electron absorbed fractions calculated from a new tomographic rat model *Phys. Med. Biol.* **53** 5343

[43] Wu L, Zhang G, Luo Q and Liu Q 2008 An image-based rat model for Monte Carlo organ dose calculations *Med. Phys.* **35** 3759–64

[44] Wang J, Fujiwara O, Wake K and Watanabe S 2008 Dosimetry evaluation for pregnant and fetus rats in a near-field exposure system of 1.95-GHz cellular phones *IEEE Microw. Wirel. Compon. Lett.* **18** 260–2

[45] Pain F, Dhenain M, Gurden H, Routier A, Lefebvre F and Mastrippolito R *et al* 2008 A method based on Monte Carlo simulations and voxelized anatomical atlases to evaluate and correct uncertainties on radiotracer accumulation quantitation in beta microprobe studies in the rat brain *Phys. Med. Biol.* **53** 5385

[46] Zhang G, Xie T, Bosmans H and Liu Q 2009 Development of a rat computational phantom using boundary representation method for Monte Carlo simulation in radiological imaging *Proc. IEEE* **97** 2006–14

[47] Arima T, Watanabe H, Wake K, Masuda H, Watanabe S and Taki M *et al* 2011 local exposure system for rats head using a figure-8 loop antenna in 1500-mhz band *IEEE Trans. Biomed. Eng.* **58** 2740–7

[48] Larsson E, Ljungberg M, Mårtensson L, Nilsson R, Tennvall J and Strand S-E *et al* 2012 Use of Monte Carlo simulations with a realistic rat phantom for examining the correlation between hematopoietic system response and red marrow absorbed dose in Brown Norway rats undergoing radionuclide therapy with ^{177}Lu-and ^{90}Y-BR96 mAbs *Med. Phys.* **39** 4434–43

[49] Xie T, Liu Q and Zaidi H 2012 Evaluation of S-values and dose distributions for ^{90}Y, ^{131}I, ^{166}Ho, and ^{188}Re in seven lobes of the rat liver *Med. Phys.* **39** 1462–72

[50] Xie T and Zaidi H 2013 Age-dependent small-animal internal radiation dosimetry *Mol. Imaging* **12** 364–75

[51] Xie T and Zaidi H 2013 Effect of emaciation and obesity on small-animal internal radiation dosimetry for positron-emitting radionuclides *Eur. J. Nucl. Med. Mol. Imaging* **40** 1748–59

[52] Caffrey E and Higley K 2013 Creation of a voxel phantom of the ICRP reference crab *J. Environ. Radioact.* **120** 14–8

[53] Hess C A 2014 Monte Carlo derived absorbed fractions in a voxelized model of a rainbow trout *Msc Thesis* Oregon State University

[54] Martinez N, Johnson T, Capello K and Pinder J 2014 Development and comparison of computational models for estimation of absorbed organ radiation dose in rainbow trout (Oncorhynchus mykiss) from uptake of iodine-131 *J. Environ. Radioact.* **138** 50–9

[55] Martinez N, Johnson T and Pinder J 2016 Application of computational models to estimate organ radiation dose in rainbow trout from uptake of molybdenum-99 with comparison to iodine-131 *J. Environ. Radioact.* **151** 468–79

[56] Kinase S 2008 Voxel-based frog phantom for internal dose evaluation *J. Nucl. Sci. Technol.* **45** 1049–52

[57] Caffrey E A 2012 *Improvements in the Dosimetric Models of Selected Benthic Organisms* (Corvallis, OR: Oregon State University)

[58] Padilla L, Lee C, Milner R, Shahlaee A and Bolch W E 2008 Canine anatomic phantom for preclinical dosimetry in internal emitter therapy *J. Nucl. Med.* **49** 446–52

[59] Kramer G H, Capello K, Strocchi S, Bearrs B, Leung K and Martinez N 2012 The HML's new voxel phantoms: Two human males, one human female, and two male canines *Health Phys.* **103** 802–7

[60] Stabin M G, Kost S D, Segars W P and Guilmette R A 2015 Two realistic beagle models for dose assessment *Health Phys.* **109** 198–204

[61] Gomez Fernandez M E 2016 Creation and Application of Voxelized Dosimetric Models: An Evaluation of Dose Conversion Factor in Apis Mellifera *Msc Thesis* Oregon State University

[62] Toivonen T, Toivo T, Pitkäaho R, Puranen L, Silfverhuth M and Mennander A *et al* 2008 Setup and dosimetry for exposing anaesthetised pigs *in vivo* to 900 MHz GSM mobile phone fields *Bioelectromagnetics* **29** 363–70

[63] Wake K, Hongo H, Watanabe S, Taki M, Kamimura Y and Yamanaka Y *et al* 2007 Development of a 2.45-GHz local exposure system for in vivo study on ocular effects *IEEE Trans. Microw. Theory Tech.* **55** 588–96

[64] Caffrey E, Johansen M and Higley K 2016 Voxel modeling of rabbits for use in radiological dose rate calculations *J. Environ. Radioact.* **151** 480–6

[65] Xie T and Zaidi H 2016 Development of computational small animal models and their applications in preclinical imaging and therapy research *Med. Phys.* **43** 111–31

[66] Zaidi H and El Naqa I 2010 PET-guided delineation of radiation therapy treatment volumes: A survey of image segmentation techniques *Eur. J. Nucl. Med. Mol. Imaging* **37** 2165–87

[67] Boone J M, Velazquez O and Cherry S R 2004 Small-animal X-ray dose from micro-CT *Mol. Imaging* **3** 149–58

[68] Badea C T, Drangova M, Holdsworth D W and Johnson G A 2008 In vivo small-animal imaging using micro-CT and digital subtraction angiography *Phys. Med. Biol.* **53** R319–50

[69] Carlson S K, Classic K L, Bender C E and Russell S J 2007 Small animal absorbed radiation dose from serial micro-computed tomography imaging *Mol. Imaging Biol.* **9** 78–82

[70] Ford N L, Thornton M M and Holdsworth D W 2003 Fundamental image quality limits for microcomputed tomography in small animals *Med. Phys.* **30** 2869–77

[71] Taschereau R, Chow P L and Chatziioannou A F 2006 Monte Carlo simulations of dose from microCT imaging procedures in a realistic mouse phantom *Med. Phys.* **33** 216–24

[72] Badea C T, Fubara B, Hedlund L W and Johnson G A 2005 4-D micro-CT of the mouse heart *Mol. Imaging* **4** 110–6

[73] Wang H, Stout D and Chatziioannou A 2015 A deformable atlas of the laboratory mouse *Mol. Imag. Biol.* **17** 18–28

[74] Ruedig E, Caffrey E, Hess C and Higley K 2014 Monte Carlo derived absorbed fractions for a voxelized model of Oncorhynchus mykiss, a rainbow trout *Radiat. Environ. Biophys.* **53** 581–7

[75] Xie T and Zaidi H 2015 Construction of pregnant female phantoms at different gestation periods for radiation dosimetry *Proc. of the 5th International Workshop on Computational Phantoms for Radiation Protection, Imaging and Radiotherapy; Seoul, Korea, 19-22 July 2015*

[76] Zaidi H 1999 Relevance of accurate Monte Carlo modeling in nuclear medical imaging *Med. Phys.* **26** 574–608

[77] Kawrakow I and Rogers D The EGSnrc code system: Monte Carlo simulation of electron and photon transport http://irs.inms.nrc.ca/software/egsnrc/

[78] Briesmeister J F 1986 MCNP-A general Monte Carlo code for neutron and photon transport (Los Alamos Radiation Transport Group, USA) *LA-7396-M*

[79] Agostinelli S, Allison J, Ka A, Apostolakis J, Araujo H and Arce P *et al* 2003 GEANT4—a simulation toolkit *Nucl. Instrum. Methods Phys. Res.* A **506** 250–303

[80] Zhu C and Liu Q 2013 Review of Monte Carlo modeling of light transport in tissues *J. Biomed. Opt.* **18** 050902

[81] Remcom 2015 http://remcom.com

[82] Oskooi A F, Roundy D, Ibanescu M, Bermel P, Joannopoulos J D and Johnson S G 2010 MEEP: A flexible free-software package for electromagnetic simulations by the FDTD method *Comput. Phys. Commun.* **181** 687–702

[83] 2009 Studio CM. Computer simulation technology (Darmstadt)

[84] Yeom Y S, Jeong J H, Han M C and Kim C H 2014 Tetrahedral-mesh-based computational human phantom for fast Monte Carlo dose calculations *Phys. Med. Biol.* **59** 3173–85

[85] Han M C, Kim C H, Jeong J H, Yeom Y S, Kim S and Wilson P P *et al* 2013 DagSolid: A new GEANT4 solid class for fast simulation in polygon-mesh geometry *Phys. Med. Biol.* **58** 4595–609

[86] Mauxion T, Barbet J, Suhard J, Pouget J-P, Poirot M and Bardies M 2013 Improved realism of hybrid mouse models may not be sufficient to generate reference dosimetric data *Med. Phys.* **40**

IOP Publishing

Computational Anatomical Animal Models
Methodological developments and research applications
Habib Zaidi

Chapter 2

Design and construction of computational animal models

Tianwu Xie, Habib Zaidi and Paul Segars

2.1 Introduction

Computational models, or phantoms, have been used extensively in medical imaging physics research and radiation dosimetry calculations where Monte Carlo- or numerical-based modeling are integrated with them to simulate the transport of ionizing or non-ionizing radiation within living organisms or imaging systems [1–4]. Such phantoms are constructed using solid-geometry shapes to represent the exterior and interior anatomical features of a given subject. The shapes defining a model can be assigned multiple physicochemical and material characteristics, such as tissue density, chemical composition, electric conductivity, relative permittivity, and scattering and absorption coefficients for different types of radiation depending on the targeted application. Until 2004, two major categories of computational animal phantoms were available, namely mathematical models and voxel-based models. Mathematical or stylized models use simple equation-based geometries such as spheres, cylinders and ellipsoids to represent organs or tissues, whereas voxel-based phantoms adopt three-dimensional cubic voxels having different physical characteristics, such as density and chemical compositions, for object representation. Starting in 2000, a third type of computational phantom emerged, referred to as boundary representation (BREP) models. Utilizing more powerful computerized 3D modeling technologies, they combine the advantages of mathematical and voxel-based models. BREP models adopted advanced surface representations such as non-uniform rational B-spline surfaces (NURBS) and 3D polygon meshes that can realistically model the tissues and organs as well as provide the flexibility to simulate anatomical variations and motion. BREP phantoms can also be converted back to voxel models for implementing radiation transport calculations with Monte Carlo- or numerical-based simulation codes. BREP models allow users to manipulate internal

doi:10.1088/2053-2563/aae1b4ch2

organs and body contours, refining selected regions as desired and generating customized computational phantoms for different applications.

Using the above techniques, a non-negligible number of computational human phantoms have been reported and used in radiation dosimetry and medical imaging simulations [5–15]. The number of reported animal models is significantly less and was mainly motivated by the need for radiation dose estimation in preclinical studies and radiation protection in radioecology research. During the last few decades, with the popularity of laboratory animals in preclinical research, different computational animal models have been developed.

In this chapter, we discuss the techniques that have been used to develop computational animal models under the three main categories mentioned above: mathematical, voxel-based, and BREP phantoms. Each type of modeling has its advantages and disadvantages. The computational efficiency and geometrical compatibility of the computational phantoms and simulation tools are important considerations in deciding which to use for a given study. Simple geometries are useful to investigate fundamental issues such as scatter and attenuation, but more complex models, including realistic organ shapes and detailed internal structures, are needed to simulate clinically realistic conditions to evaluate organ dose or image quality. Therefore, for a given application, it is essential to design a computer phantom so that results produced using it are representative of reality.

2.2 Mathematical phantoms

Mathematical or stylized phantoms define the organs and structures for a given animal using analytical equations or simple geometric primitives. The size, shape, and orientation for the geometrical primitives (ellipsoids, cones, cylinders, rectangular volumes, etc) defining each anatomical structure can be estimated from medical images or animal dissections. Overlap, cut planes, intersections, and Boolean operations can be applied to the geometric objects to form more complicated anatomical shapes. Figure 2.1 shows an example of this for the 4D Mathematical Cardiac-Torso (MCAT) phantom [16, 17], a human model, including the cardiac and respiratory motions, used in imaging research. Similar techniques have been applied to create many different animal models [18]; figure 2.2 shows two examples of mathematical phantoms developed for rodents. Different shapes can be seen to model the various structures in both.

In mathematical phantoms, the body and all internal organs are represented by a series of simplified functions which are computationally efficient and geometrically compatible with most simulation tools. Since they are defined analytically, exact ray-tracing calculations can be performed on these phantoms enabling faster simulations. Mathematical phantoms are also very flexible; they can be easily manipulated by altering the equations or the parameters of the geometric shapes that define them in order to create anatomical variations (different ages, body or organ sizes, or organ shapes, etc) or to model organ motion (voluntary or involuntary), such as that done within the MCAT. However, the simplicity with which the phantoms are defined limits their ability to realistically model the anatomy. Combinations of simple

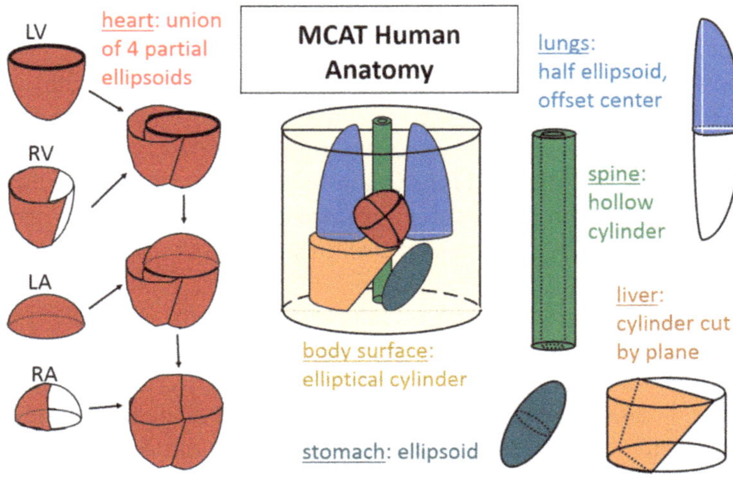

Figure 2.1. Human anatomy as modeled by the MCAT phantom using simple geometric primitives sculpted into anatomical shapes using Boolean operations, cut planes, and intersections.

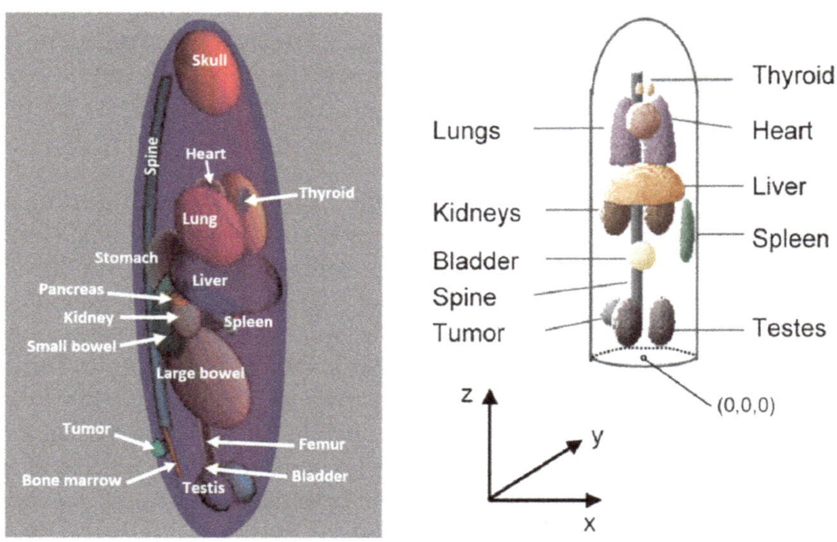

Figure 2.2. Examples of mathematical animal phantoms. (Left) rat model developed by Konijnenberg *et al* [19]. (Right) mouse model developed by Hindorf *et al* [20].

geometric primitives cannot fully model the complex anatomical shapes in the body. With their more stylized representations of the anatomy, mathematical phantoms are commonly referred to as stylized phantoms.

2.3 Voxel-based phantoms

Voxel-based or voxelized phantoms are based on segmented imaging data and therefore, provide much more realistic models of anatomy as compared to mathematical phantoms. Computed tomography (CT), magnetic resonance images

(MRI), positron emission tomography (PET), high resolution cryosection images, and other tomographic imaging data have been used to construct volume segmented models. Voxel-based phantoms are created from such data by segmenting the organs and structures of the body assigning unique index values to them.

The segmentation process to define a voxelixed phantom can be performed manually. Using available segmentation software, the operator goes through the image slices of a dataset and manually contours the internal and external structures filling them in with unique values. Figure 2.3 shows the manual segmentation of mouse CT data using the ImageSegment program developed by Segars *et al* [21]. The different colors in the image slices indicate the segmented structures. Manual segmentation such as this is demanding, requires excellent knowledge of the anatomy, and can take a significant amount of time, on the order of several weeks to a year. A number of supervised and unsupervised automated segmentation techniques have been used to alleviate the task of manual segmentation using image processing software for tissue identification (e.g. 3D-Doctor, Seg3D, ITK-Snap, and ImageJ). This includes thresholding, variational approaches, statistical learning methods, and techniques based on stochastic modeling [22]. In general, these techniques perform well for segmenting the skeleton, organs, and other structures of higher contrast but are less practical for adjoining lower contrast structures having similar density and composition. Human intervention is still often needed to

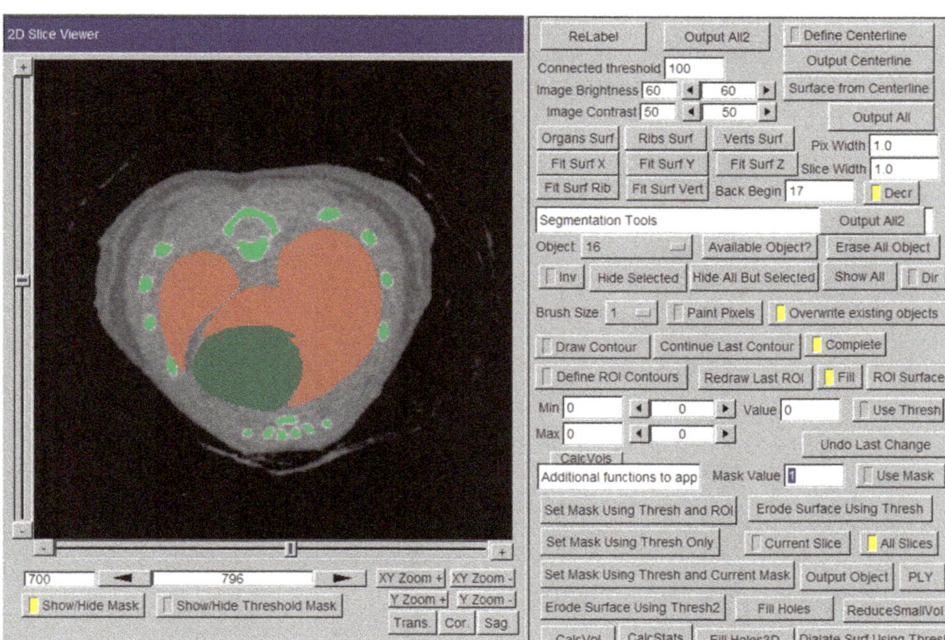

Figure 2.3. Example of manual segmentation of mouse CT data. Bones (light green), liver (brown), and heart (green) are manually contoured in the transaxial image slice and filled in with unique values corresponding to their displayed colors.

manually draw and correct boundaries in order to accurately designate some internal structures.

Recent work has focused on the development of deep learning algorithms for fully automatic image segmentation [23]. Such algorithms utilize multi-layered neural networks that are trained to segment individual organs based on sets of pre-segmented data, manually delineated by physicians or other such experts. The success of these methods is dependent on the quality and size of the training data. With manual segmentation being a time-consuming process, obtaining such a large amount of high-quality pre-segmented data for training is problematic. With proper development and training, however, deep learning techniques are obtaining very promising results.

Through the segmentation of imaging data, many different voxel-based animal models have been developed for research. The article by Xie *et al* [18] details many of these models. Figure 2.4 shows an example of a voxelized model created for a Wistar rat. A segmented image slice is shown as well as 3D renderings of the body.

Voxel-based phantoms offer much improved realism over their mathematical counterparts, but they do have their disadvantages. As can be seen in figure 2.4, object boundaries of voxelized models are not smooth and continuous as they are in mathematical models; they have stair-cased boundaries due to the slice thickness of the images upon which they were derived. Higher resolution images can be used to alleviate this, but such large datasets would require a great deal of memory and may be difficult to obtain. Fixed to a particular image dataset, the rigid structure of the voxel-based phantom also makes it difficult to adjust or deform the anatomical structures in order to model anatomical variations or motion. In addition, voxelized phantoms are defined at the particular resolution of the images upon which they are based; generation at other resolutions requires interpolation, which induces error.

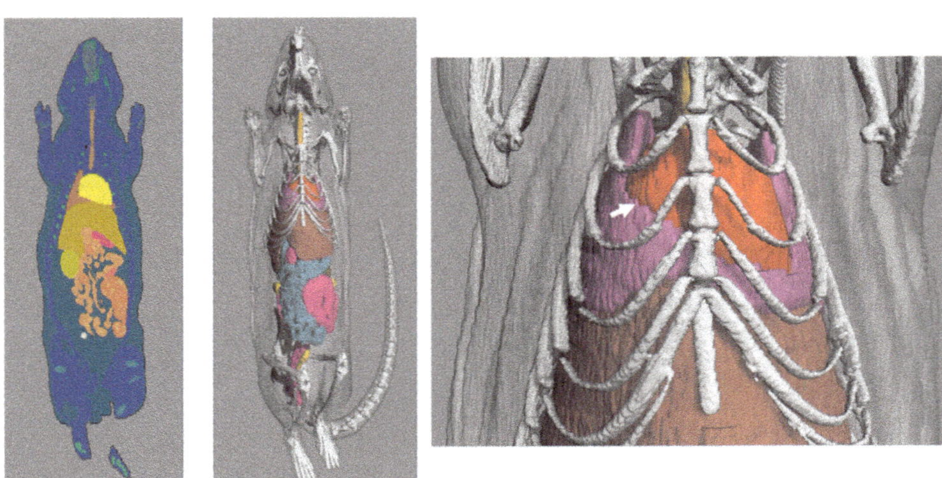

Figure 2.4. Example of a voxelized phantom of a female Wistar rat. A coronal slice from the segmented data is shown at the left while 3D renderings of the phantom are shown to the middle and right. The white arrow in the rightmost image points out the stair-casing that occurs at object boundaries due to the voxelized nature of the models.

Also, exact ray-tracing calculations are difficult and time-consuming since the organ shapes are not analytically defined.

2.4 BREP phantoms

BREP or hybrid phantoms are the latest technique in model development seeking to combine the advantages of voxel-based and mathematical phantoms. BREP models initially start off as voxel-based phantoms in that imaging data is first segmented to define the internal and external structures. The segmented structures are then fit with continuous, smooth surfaces using 3D modeling software such as Autodesk 3ds Max and Rhinoceros. B-splines, NURBS and polygon meshes have been typically used to define the surfaces for the anatomical objects. Figure 2.5 illustrates the surface fitting process to model the kidney.

Based on imaging data, BREP phantoms can accurately model each structure in the body providing the realism of a voxelized model. They also provide the flexibility of a mathematical phantom in that they can be easily modified to model anatomical variations and motion. Spline surfaces can be deformed through the control points which define their shape while polygon meshes can be altered through their vertex points. With the characteristics of both voxel-based and mathematical models, BREP phantoms are commonly referred to as hybrid phantoms. Different BREP or hybrid phantoms have been developed recently including models for dogs, mice, and rats [18]. Figure 2.6 shows the first hybrid mouse model, the mouse whole-body (MOBY) phantom developed by Segars *et al* [24].

Being defined with more complex surfaces, BREP phantoms cannot be directly adopted within defined geometries of most simulation tools. BREP phantoms are typically rendered back into a 3D voxelized format, with the organs set to index values or user-defined property values, in order to operate within existing analytical or Monte Carlo based simulation packages. Since the organs and structures are defined by continuous, smooth surfaces, the phantoms can be generated at any

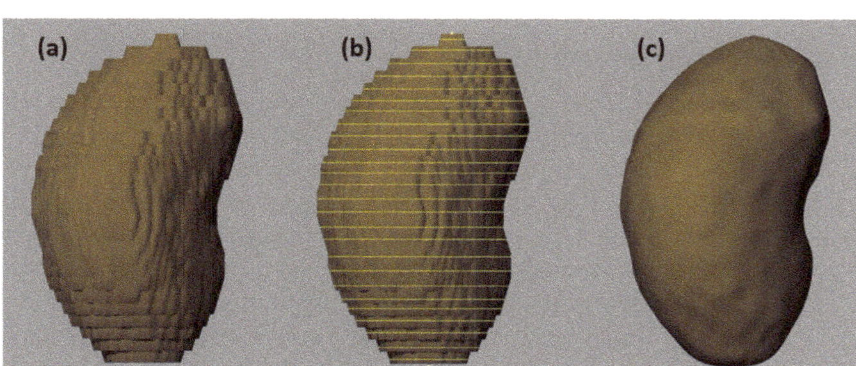

Figure 2.5. BREP modeling of the kidney. Kidney segmented from imaging data is imported into the Rhinoceros modeling software (a). Contours are taken from the segmented model and the starting and ending points are defined (b). A cubic NURBS surface is then lofted through the contours, starting and ending at the defined points. The NURBS surface provides a smoother representation of the kidney.

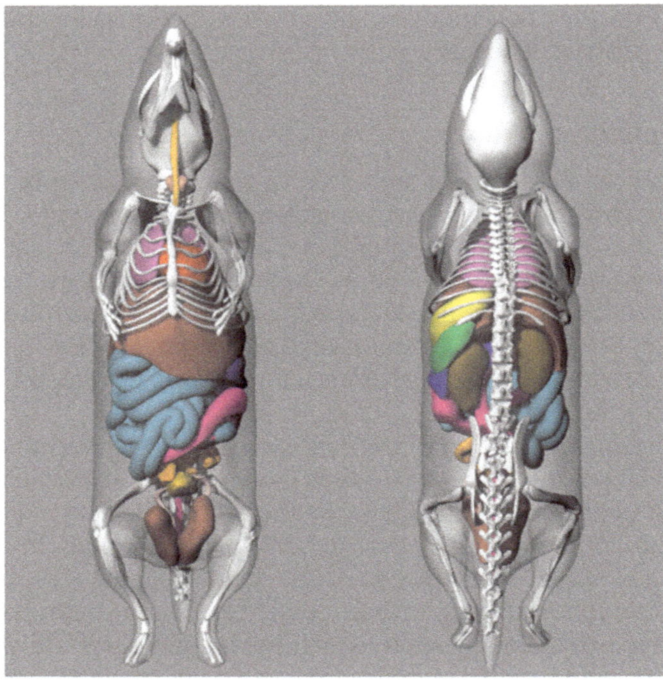

Figure 2.6. MOBY phantom modeling the anatomy of a mouse. Anterior (left) and posterior (right) views are shown. Organs and structures are modeled with NURBS surfaces. Using the flexibility of the NURBS definition, the MOBY phantom includes models for the cardiac and respiratory motions.

resolution without using interpolation. Direct ray-tracing calculations are possible from hybrid phantoms, although they may require additional computational time. Ray-tracing calculations of polygon meshes are straightforward and can be done analytically. NURBS surfaces, on the other hand, involve iterative procedures which are more intensive.

2.5 Summary and future perspectives

In summary, three main techniques have been used to construct animal models for research: mathematical, voxel-based, and BREP. Mathematical or stylized models are based on simple geometric primitives or equations while voxelized models are derived from the segmentation of tomographic image data. More sophisticated BREP or hybrid models start off as voxel-based phantoms but go a step beyond by fitting advanced surface representations to the segmented structures. This gives them the flexibility advantageous of mathematical phantoms to model anatomical variations or motion. Figure 2.7 compares the realism and degree of geometric sophistication of mathematical, voxel, and BREP models developed based on the same rat dataset [25]. In designing a computational animal phantom, it is important to determine the level of realism required for the target application. Simpler geometry-based phantoms may be useful for some tasks while more sophisticated voxelized or hybrid models are needed for others.

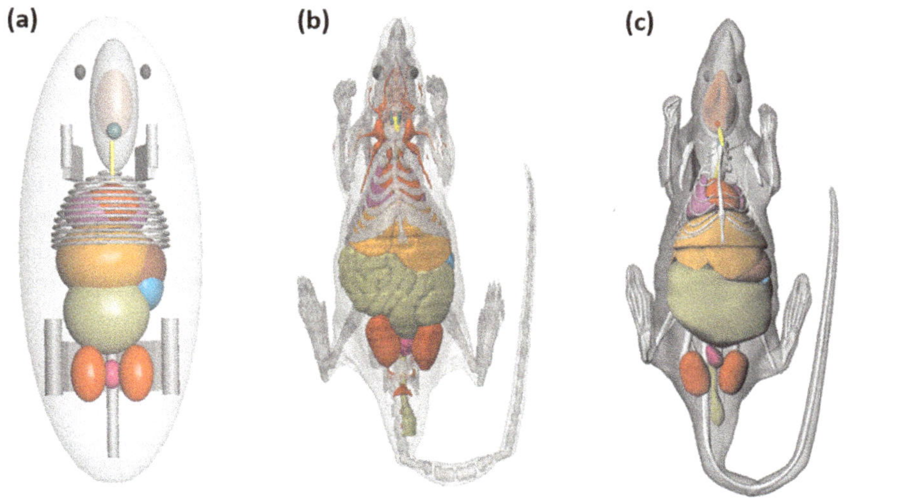

Figure 2.7. Anterior view of a (a) mathematical, (b) voxelized, and (c) hybrid rat phantom derived from the same imaging dataset. Reprinted with permission from [25]. Copyright 2010, AIP Publishing LLC.

Looking forward, we believe the following aspects of computational animal models should be further investigated:

(1) Cross-comparison of voxel-based models derived from different hybrid models. A number of groups have customized their animal models based on hybrid models [26–33]. Organ shapes and locations of generated voxel models are similar to the original hybrid models. However, the impact of the selection of the original hybrid model on the dosimetric differences between customized models is not well understood. A systematic cross-comparison of customized compatible animal models and quantification of the dosimetric differences related to different original hybrid models is needed.

(2) A shift away from physical parameters of reference humans. In most published studies, the chemical composition and density of animal tissues is assumed to be the same as the reference data of humans published by the ICRP [34]. However, physical variations associated with different species can cause differences in the dose estimation [35]. The use of physical parameters of reference humans for small animal radiation dosimetry may introduce uncertainty. Future radiation dosimetry calculations of laboratory animals used in preclinical research may require this uncertainty to be quantified and reduced. To achieve that future goal, a survey of basic anatomical and physiological data for common laboratory animals is highly desired.

(3) Advanced simulation tools for hybrid models. Most Monte Carlo tools for radiation transport simulations were designed for nuclear applications and suffer from poor software engineering design for handling complex geometries. As such, most hybrid models need to be converted to voxel models to enable their implementation in simulation tools. Further research and development efforts are required to develop advanced simulation tools

enabling the handling of hybrid geometries for radiation transport calculations and dynamic objects, such as moving lungs and beating heart.

As further developments are made and more phantoms are developed, the use of computational animal models for research will continue to rise. They provide the tools necessary to develop, quantitatively evaluate, and improve imaging devices and techniques and to investigate the effects of anatomy and motion. They are also instrumental in radiation dosimetry to estimate dose and radiation risk and optimize dose reduction strategies, an important area of research given the high amounts of radiation exposure attributed to medical imaging procedures.

References

[1] Xu X G 2014 An exponential growth of computational phantom research in radiation protection, imaging, and radiotherapy: A review of the fifty-year history *Phys. Med. Biol.* **59** R233–302

[2] Xu X G and Eckerman K F 2010 *Handbook of Anatomical Models for Radiation Dosimetry* (Boca Raton, FL: CRC Press)

[3] Zaidi H and Tsui B M 2009 Review of computational anthropomorphic anatomical and physiological models *Proc. IEEE* **97** 1938–53

[4] Zaidi H and Xu X G 2007 Computational anthropomorphic models of the human anatomy: The path to realistic Monte Carlo modeling in radiological sciences *Annu. Rev. Biomed. Eng.* **9** 471–500

[5] Caon M 2004 Voxel-based computational models of real human anatomy: A review *Radiat. Environ. Biophys.* **42** 229–35

[6] Cassola V, de Melo Lima V, Kramer R and Khoury H 2009 FASH and MASH: Female and male adult human phantoms based on polygon mesh surfaces: I. Development of the anatomy *Phys. Med. Biol.* **55** 133

[7] Lee C, Lodwick D, Hurtado J, Pafundi D, Williams J L and Bolch W E 2009 The UF family of reference hybrid phantoms for computational radiation dosimetry *Phys. Med. Biol.* **55** 339

[8] Petoussi-Henss N, Zankl M, Fill U and Regulla D 2001 The GSF family of voxel phantoms *Phys. Med. Biol.* **47** 89

[9] Xie T, Bolch W E, Lee C and Zaidi H 2013 Pediatric radiation dosimetry for positron-emitting radionuclides using anthropomorphic phantoms *Med. Phys.* **40** 102502–14

[10] Xie T, Kuster N and Zaidi H 2017 Computational hybrid anthropometric paediatric phantom library for internal radiation dosimetry *Phys. Med. Biol.* **62** 3263–83

[11] Xie T and Zaidi H 2014 Effect of respiratory motion on internal radiation dosimetry *Med. Phys.* **41** 112506–10

[12] Xie T and Zaidi H 2014 Evaluation of radiation dose to anthropomorphic paediatric models from positron-emitting labelled tracers *Phys. Med. Biol.* **59** 1165–87

[13] Xie T and Zaidi H 2014 Fetal and maternal absorbed dose estimates for positron-emitting molecular imaging probes *J. Nucl. Med.* **55** 1459–166

[14] Xie T and Zaidi H 2015 Construction of pregnant female phantoms at different gestation periods for radiation dosimetry *Proc. of the 5th Int. Workshop on Computational Phantoms for Radiation Protection, Imaging and Radiotherapy; Seoul, Korea, pp 19–22*

[15] Xie T and Zaidi H 2016 Development of computational pregnant female and fetus models and assessment of radiation dose from positron-emitting tracers *Eur. J. Nucl. Med. Mol. Imaging* **43** 2290–300

[16] Pretorius P H, King M A, Tsui B M, LaCroix K J and Xia W 1999 A mathematical model of motion of the heart for use in generating source and attenuation maps for simulating emission imaging *Med. Phys.* **26** 2323–32
[17] Segars W P, Lalush D S and Tsui B M W 2001 Modeling respiratory mechanics in the MCAT and spline-based MCAT phantoms *IEEE Trans. Nucl. Sci.* **48** 89–97
[18] Xie T and Zaidi H 2016 Development of computational small animal models and their applications in preclinical imaging and therapy research *Med. Phys.* **43** 111
[19] Konijnenberg M W, Bijster M, Krenning E P and De Jong M 2004 A stylized computational model of the rat for organ dosimetry in support of preclinical evaluations of peptide receptor radionuclide therapy with (90)Y, (111)In, or (177)Lu *J. Nucl. Med.* **45** 1260–9
[20] Hindorf C, Ljungberg M and Strand S E 2004 Evaluation of parameters influencing S values in mouse dosimetry *J. Nucl. Med.* **45** 1960–5
[21] Segars W P, Sturgeon G, Mendonca S, Grimes J and Tsui B M W 2010 4D XCAT phantom for multimodality imaging research *Med. Phys.* **37** 4902–15
[22] Zaidi H and El Naqa I 2010 PET-guided delineation of radiation therapy treatment volumes: A survey of image segmentation techniques *Eur. J. Nucl. Med. Mol. Imaging* **37** 2165–87
[23] Litjens G, Kooi T, Bejnordi B E, Setio A A A, Ciompi F and Ghafoorian M *et al* 2017 A survey on deep learning in medical image analysis *Med. Image Anal.* **42** 60–88
[24] Segars W P, Tsui B M W, Frey E C, Johnson G A and Berr S S 2004 Development of a 4-D digital mouse phantom for molecular imaging research *Mol. Imaging Biol.* **6** 149–59
[25] Xie T W, Zhang G Z, Li Y and Liu Q A 2010 Comparison of absorbed fractions of electrons and photons using three kinds of computational phantoms of rat *Appl. Phys. Lett.* **97** 3
[26] Keenan M A, Stabin M G, Segars W P and Fernald M J 2010 RADAR realistic animal model series for dose assessment *J. Nucl. Med.* **51** 471–6
[27] Larsson E, Ljungberg M, Strand S-E and Jönsson B-A 2011 Monte Carlo calculations of absorbed doses in tumours using a modified MOBY mouse phantom for pre-clinical dosimetry studies *Acta Oncol.* **50** 973–80
[28] McIntosh R L, Deppeler L, Oliva M, Parente J, Tambuwala F and Turner S *et al* 2010 Comparison of radiofrequency exposure of a mouse dam and foetuses at 900 MHz *Phys. Med. Biol.* **55** N111
[29] Mohammadi A and Kinase S 2011 Influence of voxel size on specific absorbed fractions and S-values in a mouse voxel phantom *Radiat. Prot. Dosim.* **143** 258–63
[30] Wang H, Stout D and Chatziioannou A 2015 A deformable atlas of the laboratory mouse *Mol Imag Biol.* **17** 18–28
[31] Xie T and Zaidi H 2013 Age-dependent small-animal internal radiation dosimetry *Mol. Imaging.* **12** 364–75
[32] Xie T and Zaidi H 2013 Effect of emaciation and obesity on small animal internal radiation dosimetry for positron-emitting radionuclides *Eur. J. Nucl. Med. Mol. Imaging (in press)*
[33] Xie T and Zaidi H 2013 Monte Carlo-based evaluation of S-values in mouse models for positron-emitting radionuclides *Phys. Med. Biol.* **58** 169
[34] ICRP Publication 89 2002 Basic anatomical and physiological data for use in radiological protection: reference values *Ann. ICRP* **32** 1–277
[35] Xie T, Liu Q and Zaidi H 2012 Evaluation of S-values and dose distributions for (90)Y, (131)I, (166)Ho, and (188)Re in seven lobes of the rat liver *Med. Phys.* **39** 1462–72

IOP Publishing

Computational Anatomical Animal Models
Methodological developments and research applications
Habib Zaidi

Chapter 3

Overview of computational mouse models

Akram Mohammadi, Sakae Kinase and Mitra Safavi-Naeini

3.1 Introduction

Over the last few decades, there has been great interest in using laboratory mice in preclinical experiments to evaluate the potential of new radiopharmaceuticals for the diagnosis and treatment of diseases such as cancers, to assess molecular imaging probes, to investigate the effect of ionizing and nonionizing radiation on living organisms, and to develop new medical imaging technologies and instruments. The motive for this research is the high likelihood of successful translation of results obtained with mice to humans for some or all of these applications. Different types of simulation tools have been developed to simulate radiation transport, including photons (x-rays, visible light, infrared and microwave radiation), electrons and other particles in realistic computational animal models. These enable the simulation of various medical imaging modalities, including bioluminescence tomography (BLT), fluorescence molecular tomography (FMT), single photon emission computed tomography (SPECT), positron emission tomography (PET), x-ray computed tomography (CT), and magnetic resonance imaging (MRI) [1, 2]. The simulation results may be used to predict the performance of novel designs of medical imaging instruments before investing in the construction of expensive prototypes, to provide information on energy deposited in biological tissues for both ionizing and nonionizing radiation dosimetry, and to provide simulated datasets for developing new image reconstruction algorithms for various medical imaging devices. Accurate anatomical and physiological modeling of laboratory mice is essential for such simulations for all applications [1, 2].

Computational mouse models can be defined by using three different methods: 1. equation-based mathematical functions (stylized models); 2. digital volume arrays (voxel-based models); and 3. boundary representation methods (BREP models). Stylized models represent the internal organs and the exterior contour using simplified mathematical equations, such as planes, cylinders, cones, ellipsoids, tori, and spheres. Voxel-based models use matrices of segmented voxels based on tomographic images. In BREP computational models, the surface contours of organs and tissues are

doi:10.1088/2053-2563/aae1b4ch3

described by polygon mesh or nonuniform rational B-Spline (NURBS). The stylized models are easy, simple and fast for calculations in simulations and analytic modeling work, but cannot accurately describe complex anatomic details. Better anatomic correctness is achievable with voxel-based models, but intrinsic discretization errors are unavoidable. It is also difficult to make any changes to the shape of internal organs and external contours in the voxel models, which usually limits their use to 3D applications. BREP models represent accurate anatomic features which are also deformable, therefore allowing the generation of a population of individuals with variable physical dimensions, or 4D models which incorporate breathing lungs and a beating heart. However, purely BREP models cannot be directly used in most Monte Carlo simulation tools because of difficulties encountered during the calculation of intersections of particle trajectories with surfaces [2].

This overview was written based on all computational mouse models found in well-established scientific resources and sorted in chronological order. Table 3.1 presents a summary of mouse computational models developed since 1994. The *Anatomical features* column describes the type of mouse represented by the model (such as sex and weight), *Model number* indicates the number of unique mouse models implemented by the model according to the published reference, and *Imaging modality* refers to the imaging technique which was used to derive the model. The development of 26 computational mouse models within a short time indicates the importance of the models in preclinical research and it suggests the following questions in readers' minds.

- What is the degree of realism of computational mouse models?
- How close are the models to the anatomy and physiology of laboratory mice?
- What is the effect of the choice of computational mouse model on the simulation results?
- What are the future directions?

The answers to the above questions require an understanding of the rationales and simulation techniques used for the development of computational mouse models in various applications. The review summarizes the history of development of the major computational mouse models during the last two decades with some provision about what to expect in the near future for their application in preclinical research, radiation dosimetry calculations and imaging physics research.

3.2 Construction of computational mouse models

Solid-geometry shapes are the fundamental building blocks from which computational models are constructed and are used to represent all internal organs and the body of the simulated animal. It is possible to assign several physicochemical characteristics of the anatomical features, such as tissue density, chemical composition, electric conductivity, relative permittivity, and scattering and absorption coefficients for different types of radiation according to the targeted simulation study. The construction of computational models must also consider the geometrical

Table 3.1. Summary of various mouse models [3–29].

Year	Developer	Affiliation	Anatomical features	Model number	Imaging modality	Model type	Applications
1994	Hui *et al* [3]	Pacific Northwest Laboratory, USA	25 g nude mouse	1	Anatomic data	Stylized model	Calculation of AF for ^{90}Y
1997	Yoriyaz and Stabin [4]	Oak Ridge National Laboratory, USA	30 g mouse	1	Anatomic data	Stylized model	For photon and electron transport
1999	Kennel *et al* [5]	Oak Ridge National Laboratory, USA	30 g mouse	1	Anatomic data	Stylized model	Calculation of absorbed dose for ^{90}Y and ^{213}Bi
2001	Dhenain *et al* [6]	California Institute of Technology	Mouse embryo	1	MRI	Voxel model	Nonionizing radiation applications
2001	Flynn *et al* [7]	Royal Free and University College Medical School, UK	Mouse	1	Anatomic data	Stylized model	Calculation of organ absorbed dose for ^{131}I and ^{90}Y
2001	Wang *et al* [8]	Nagoya Institute of Technology, Japan	Mouse	1	MRI	Voxel model	Effects of 1.5 GHz digital cellular phones on mouse skin carcinogenesis
2003	Kolbert *et al* [9]	Memorial Sloan-Kettering Cancer Center, USA	25 g female athymic mouse	1	MRI	Voxel model	Calculation of *S*-values for ^{131}I, ^{153}Sm, ^{32}P, ^{188}Re, and ^{90}Y
2004	Funk *et al* [10]	University of California (San Francisco), USA	20, 30, and 40 g mice with fixed axes ratios	3	Anatomic data	Stylized model	Calculation of radiation dose to mice and rats for ^{18}F, ^{99m}Tc, ^{201}Tl, ^{111}In, ^{123}I, and ^{125}I
2004	Hindorf *et al* [11]	Lund University, Sweden	24 g mouse	1	Anatomic data	Stylized model	Calculation of *S*-values for ^{90}Y, ^{131}I, ^{111}In and ^{99m}Tc
2004	Segars *et al* [12]	John Hopkins University, USA	Male C75BL/6 mouse	1	MRI	Hybrid model	Ionizing and nonionizing radiation applications

(*Continued*)

2005	Miller *et al* [13]	University of Missouri, MO, USA	25 g nude mouse with improved bone marrow model	1	Anatomic data	Stylized model	Calculation of AFs for ^{90}Y, ^{188}Re, ^{166}Ho, ^{149}Pm, ^{64}Cu, and ^{177}Lu
2006	Sato *et al* [14]	Chiyoda Technol Corporation, Japan	Mouse	1	Anatomic data	Stylized model	Calculation of absorbed dose for ^{90}Y
2004	Stabin *et al* [15]	Vanderbilt University, TN, USA	27 g transgenic mouse	1	CT	Voxel model	AFs for electrons of (0.1–4) MeV and photons of (0.01–4) MeV
2007	Bitar *et al* [16]	INSERM, France	30 g female athymic nude mouse	1	Cryosection	Voxel model	AFs for mono-energetic photon and electron sources
2007	Dogdas *et al* [17]	University of Southern California, USA	28 g male BALB/c mouse	1	PET, CT, and cryosection	Voxel model	Ionizing radiation applications
2009	Zhang *et al* [18]	Chinese Academy of Sciences, China	25 g nude mouse with improved bone marrow model	1	Anatomic data	Stylized model	Calculation of AFs for ^{90}Y, ^{188}Re, ^{166}Ho, ^{149}Pm, ^{64}Cu, and ^{177}Lu
2010	Keenan *et al* [19]	Vanderbilt University, TN, USA	25, 30, and 35 g male C57BL/6 mice	3	Derived from MOBY (MRI)	Voxel model	Radiation dose for mono-energetic electrons, photons, ^{18}F, ^{32}P, ^{124}I, ^{90}Y, ^{111}In, and ^{177}Lu
2010	McIntosh *et al* [20]	Australian Center for Radiofrequency Bioeffects Research, Australia	30 g ICR and ddY male, 22 g female mouse, 22 g pregnant mouse, and 0.5 g mouse fetus	4	Anatomic data	Voxel model	SARs in radiofrequency dosimetry for 900 MHz plane wave
2010	Chow *et al* [21]	University of Toronto, Canada	Mouse	3	CT	Voxel model	Radiation dose from micro-CT

2011	Larsson *et al* [22]	Lund University, Sweden	22, 28, and 34 g male C57BL/6 mice with tumor model	3	Derived from MOBY (MRI)	Voxel model	AFs for mono-energetic photons and electrons. *S*-values for ^{125}I, ^{131}I, ^{111}In, ^{177}Lu, and ^{90}Y
2011	Mohammadi and Kinase [23]	JAEA, Japan	28 g male mouse	1	Derived from Digimouse (PET, CT, and cryosection)	Voxel model	Radiation dose for mono-energetic photons, electrons, ^{131}I, ^{153}Sm, ^{188}Re and ^{90}Y
2011	Zhang *et al* [24]	Beijing Institute of Radiation Medicine, China	28 g mouse	1	Derived from MOBY (MRI)	Voxel model	Radiation dose from external photon beams (0.01–10 MeV)
2013	Xie and Zaidi [25]	Geneva University Hospital, Switzerland	Male C57BL/6 mice with various body sizes	17	Derived from MOBY (MRI)	Voxel model	AFs and *S*-values for ^{11}C, ^{13}N, ^{15}O, ^{18}F, ^{64}Cu, ^{68}Ga, ^{86}Y, and ^{124}I
2013	Mauxion *et al* [26]	CRTC, France	30 g mouse	1	Derived from MOBY (MRI)	Voxel model	*S*-values and SAFs for 18F with two MC code
2016	Kostou *et al* [27]	University of Patras, Greece	22, 28, and 34 g mice	3	Derived from MOBY (MRI)	Voxel model	*S*-values for ^{18}F, ^{68}Ga, ^{131}I, ^{111}In, ^{177}Lu, ^{99m}Tc, and ^{188}Re
2017	Mendes *et al* [28]	CDTN/CNEN, Brazil	25.7 g mouse	1	Derived from Digimouse	Voxel model	Radiation dose for radiopharmaceuticals
2017	Locatelli *et al* [29]	INSERM, France	20 and 25 g (male and female) mice	4	MRI	Voxel model	Dose assessment of mice contaminated with radionuclides

compatibility of the computational models and simulation tools, and the impact on computational efficiency.

The anatomical parameters of a computational model of mouse organs are estimated from animal dissections or medical images of the animal. In a stylized model of a mouse, most organs can be depicted using appropriate ellipsoids. For instance, the geometry of the lungs of a mouse can be approximated by the union of the upper parts of two ellipsoids using Boolean operations, as shown in figure 3.1(a). All internal organs and the mouse body can be represented using the same approach, and most Monte Carlo simulation codes are computationally efficient and geometrically compatible with the models. However, significant anatomical uncertainties are introduced into the simulation because of the degree of simplification of the shapes of organs.

Voxel models using cuboids are constructed directly from medical image data, and their geometry is very easy to handle in all simulation codes. A large number of image datasets derived from high-resolution cryosection images, CT, MRI, or PET scans are required for construction of 3D voxel models. A segmentation process must be performed for each slice of image data to assign each voxel to a particular tissue of interest using a unique identification number. The manual segmentation of image data is a challenging and time-consuming process, and it requires excellent knowledge of the mouse anatomy. Segmentation also can be performed automatically using image processing software (e.g. Photoshop and ImageJ) based on stochastic modeling [30] for tissue identification. While this works well for segmentation of regions with high contrast between tissue densities and compositions, this does not apply to all internal organs. Therefore, even with automatic image segmentation, some manual input is usually necessary.

In voxel models, an organ has a step-shaped boundary instead of the smooth surface of the real organ, as shown in figure 3.1(b) for the lungs of a MOBY model. Consequently, the accuracy of the model strongly depends on the adopted voxel size, which is determined by the resolution of the medical imaging modality used to generate it. The voxel size is about 100–250 μm for micro-MRI with very high strength magnetic fields, around 50 μm for micro-CT and around 20 μm for cryosection images [1]. The voxel model represents the computational model of a specific animal with a fixed geometry of organs; therefore, adjustment of organ

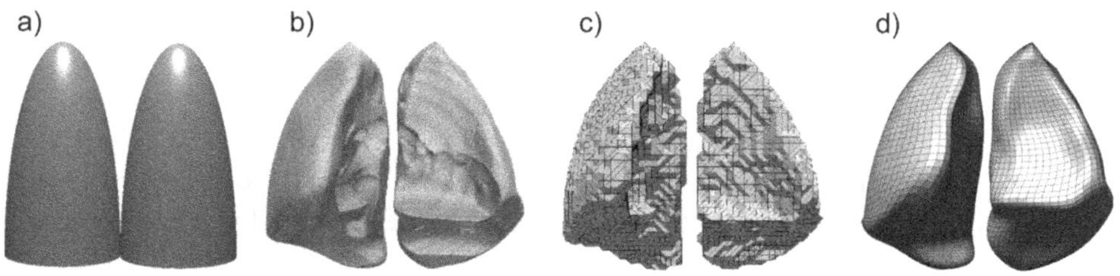

Figure 3.1. The mouse lung model defined by different models. (a) The stylized model using solid geometry. (b) The voxel-based model of the lungs. (c) The lung model presented using a polygon mesh surface. (d) The lung model represented by a NURBS surface.

shape or deformation of the model (for example, to account for breathing or heart beating in a 4D model) is difficult.

The internal organs and body shape of a mouse in a computational model can also be defined by the advanced BREP modeling techniques involving NURBS and polygon mesh surfaces. In both techniques, surface contours defined the boundaries of all organs and the modeled organs have smooth shapes and realistic anatomy. Figure 3.1(c) shows a polygon mesh model of the MOBY lungs, rendered using ParaView 5.4.1 [31]. The NURBS model of the lungs, rendered in MATLAB, is shown in figure 3.1(d). The computational models made using mesh and/or NURBS modeling techniques are broadly referred to as 'BREP' models or hybrid models. The size and shape of the organ in the mesh or NURBS model can be easily changed or deformed because of the flexibility and parametric nature of the surface contours of the organs. However, most simulation tools are not compatible with the hybrid models due to fundamental differences in the means by which organ and simulation geometry is defined. The problem can be solved by obtaining specific 3D or 4D voxel models from the hybrid models for the simulation. The flexibility of the hybrid models for deformation and the capability of voxel models for simulation has greatly advanced the capabilities of contemporary computational models in preclinical and clinical radiation dosimetry and medical imaging research.

3.3 History of computational mouse models

Laboratory mice are widely used in different fields of scientific research, such as modeling in genetics, molecular imaging, radiation dosimetry and development of novel radiopharmaceuticals. They have been used extensively to study the parameters that affect biodistribution, pharmacokinetics and radiotoxicity of novel radiotracers. The uptake, retention and effects of the radiotracers in normal tissues as well as their effects on tumor growth have been evaluated in mice before designing human clinical trials. Radiation dosimetry is indispensable to limit the dose delivered for these laboratory animals to less than the lethal value in preclinical research. It is also important to evaluate the radiation dose received by these small animals from multimodal imaging devices (e.g. PET/CT or SPECT/CT), since the delivered dose can induce changes in biological pathways and hence impact the experimental outcomes [32]. The lethal dose (LD) is a dose limitation factor for small animals; $LD_{50/30}$ is defined as the whole-body radiation dose that would kill 50% of the animals exposed to radiation within 30 days, and it is reported to be within the range 5–7.6 Gy depending on the age of the animal at the time of exposure and many other factors [33–35]. Typical whole-body x-ray radiation dose from micro-CT imaging reported in the literature ranges from 0.02–0.8 Gy [33, 36] depending on required image quality, and it can be up to 1.5 Gy for cine cardiac micro-CT [37]. On the other hand, the radiation dose in mice from radiotracers in nuclear medicine depends on the tracer type, injected activity and the size of mouse. The reported dose in the literature for [^{18}F]FDG in a 21 g mouse is 8.9 mGy/Bq [25].

The first computational mouse model was developed by Hui *et al* [3] in 1994 to reliably estimate internal doses received by organs in a laboratory mouse from ^{90}Y labeled immunoconjugates. Hui *et al* constructed the mouse model based on organ size and proximity measurements of 10 athymic mice, and all organs were modeled as ellipsoids with the exception of bone and marrow (which were modeled as cylinders). Hereupon, various computational mouse models have been reported. Yoriyaz and Stabin [4] developed a 30 g stylized mouse model with major organs modeled as ellipsoids and Kennel *et al* [5] evaluated *S*-values and absorbed fractions for that mouse model. Another stylized mouse model with advanced kidney and tumor models was constructed by Flynn *et al* [7], whereas three simplified mouse models with fixed axis ratios were presented by Funk *et al* [10] for whole-body dose estimation. Two more mouse stylized models were also reported in the literature, including the stylized model of Hui *et al* with a range of tumor sizes [13] and a stylized mouse model composed of three small cuboids for calculation of absorbed dose for ^{90}Y [14].

A digital mouse atlas was constructed by Dhenain *et al* [6] using high-resolution MR images for nonionizing radiation application, whereas another mouse voxel model was obtained by Wang *et al* [8] by reducing the voxel size of a rat model to study the biological effects of cellular phones. Chow *et al* [21] developed three voxel models of heterogeneous, homogeneous and bone homogeneous mouse from CT images of the same animal to study the impact of photon energy and heterogeneities on dose distribution and dosimetric characteristics. A voxel model of a 25 g female athymic mouse was reconstructed for calculation of *S*-values for the liver, spleen, and kidneys by Kolbert *et al* [9], while a 28 g voxel-based mouse model from cryosection image data was reported by Zhang *et al* [18]. Several other voxel-based mouse models have also been developed, including a 27 g transgenic mouse model based on CT images [15], a 30 g female athymic nude mouse obtained from cryosection images [16], and a 20 g male BALB/c mouse [18]. In 2007, the well-known voxel-based model of Digimouse was developed by Dogdas *et al* [17] from CT, cryosection, and PET images of a 28 g nude male mouse, and the model was used by Mendes *et al* [28] and later modified by Mohammadi and Kinase [23] for dose calculation of all organs. Four voxel-based models of a 30 g male mouse, a 22 g female mouse, a 22 g pregnant mouse, and a 0.5 g mouse fetus were constructed by McIntosh *et al* [20] for evaluation of radiofrequency exposure. Recently, Locatelli *et al* [29] developed 20 g male mouse and 25 g female mouse voxel-based models using MR imaging for dose assessment of mice contaminated with radionuclides.

The first hybrid mouse model, MOBY, was built using MRI images of a male C57BL/6 mouse by Segars *et al* [12] in 2004. This model was originally developed for use in molecular imaging research, however, it has been widely used in many other research fields in addition to imaging physics and radiation dosimetry. A number of mouse models have been derived from the MOBY model, including three voxel-based models by Keenan *et al* [19], 17 voxel mouse models with various body sizes by Xie and Zaidi [25], three voxel-based tumor-bearing mouse models by Larsson *et al* [22], a 30 g voxel-based model by Mauxion *et al* [26], a 28 g voxel mouse by Zhang *et al* [24], and three voxel mouse models of 22, 28, and 34 g by Kostou *et al* [27]. A deformable mouse atlas with the capability of changing body weight and body poses was also reported in the literature [38, 39].

Figure 3.2 shows the stylized mouse model of Hindorf *et al* [11] developed for estimation of *S*-values for various radionuclides. Figures 3.3 and 3.4 show the voxel mouse model of Digimouse [17], and the first hybrid mouse model of MOBY [12].

3.4 Simulation tools used with the computational mouse models

The mouse computational models are used for characterization of the real mouse anatomy and simulation of radiation transport inside the mouse body to provide

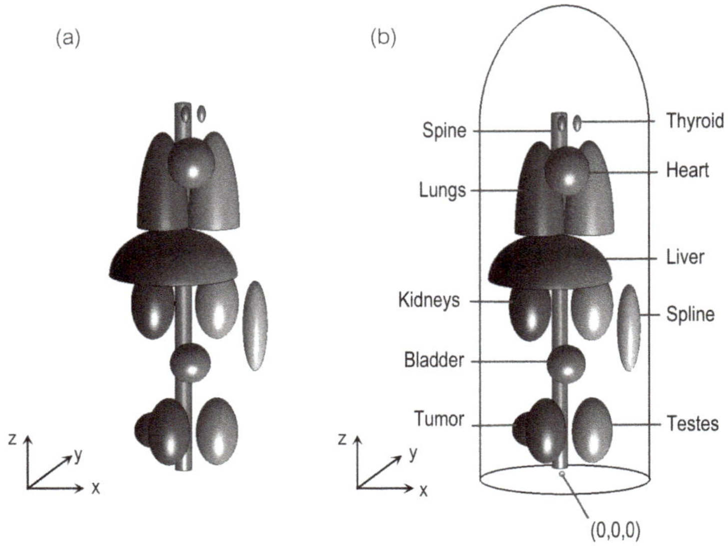

Figure 3.2. The stylized mouse model of Hindorf *et al* (a) without skin and (b) with skin.

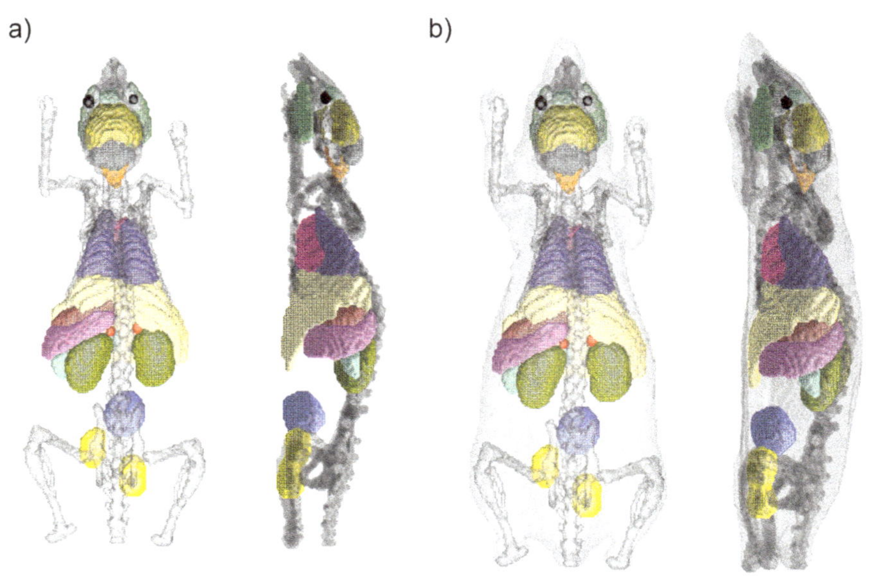

Figure 3.3. The voxel-based model of Dogdas *et al*, Digimouse, (a) without skin and (b) with skin.

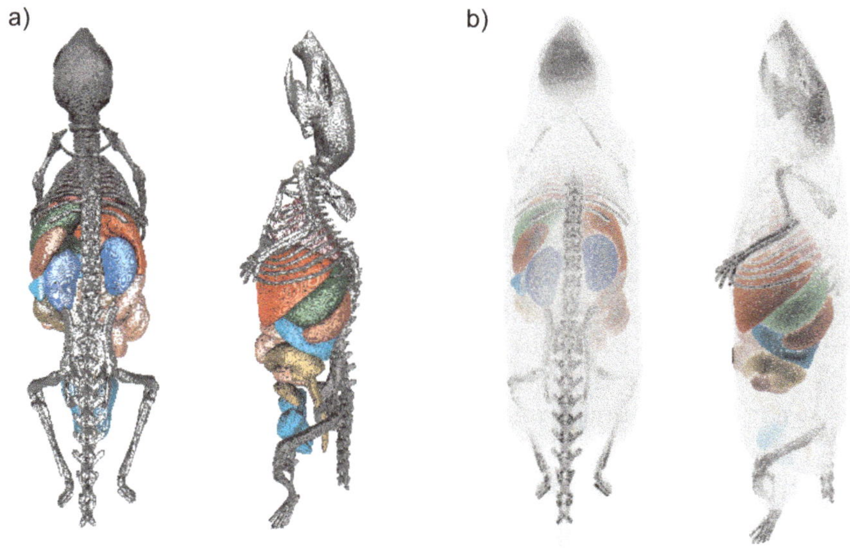

Figure 3.4. The hybrid mouse model of Segars *et al*, MOBY, (a) without skin and (b) with skin.

information on patterns of radiation interactions and distributions of deposited energy. Monte Carlo simulation codes are capable of accurate modeling of various types of transport of ionizing radiation inside the mouse body by considering all relevant aspects of particle interactions within 3D heterogeneous media [40, 41]. As a result, Monte Carlo codes are widely used for many applications in medical radiation physics, e.g. internal dosimetry of radiotracers, imaging physics research including imaging modalities such as CT, PET, and SPECT, and external beam radiotherapy applications. Some of the public-domain and general-purpose Monte Carlo codes (e.g. EGS4, MCNP, and GEANT4) are capable of simulation of the complex geometry of computational mouse models and are available as open-source software. The electron gamma shower (EGS) [42] code is a general-purpose package for Monte Carlo simulation of the coupled transport of electrons and photons in an arbitrary geometry and written in MORTRAN by a team at the Stanford Linear Accelerator Center. The Monte Carlo N-Particle (MCNP) [43] code is used for simulating particle interactions of several types at various energies and is written in FORTRAN by Los Alamos National Laboratory. The GEANT4 [44] code is a toolkit for simulating the transport of particles through matter and it was developed by the European Organization for Nuclear Research in C++. GATE is one of the popular GEANT4-based computational codes that does not need computer programming knowledge, developed by the international OpenGATE collaboration for numerical simulation in medical imaging and radiotherapy [45]. Additionally, the Monte Carlo simulation techniques have numerical solutions for light transport equations in biological tissues and are frequently used to simulate light distributions for given optical properties or to estimate optical properties for experimentally measured light distribution [46]. MOSE is a computer package for simulation of light propagation in a living mouse using Monte Carlo techniques [47].

Numerical techniques are also intensively used in computational electromagnetism because of the capability of these techniques to derive closed form solutions of Maxwell's equations. Three main numerical methods based on discretization techniques, including the method of moment (MoM) by calculating only boundary values, the finite-element method (FEM) by applying unstructured grids, and the finite-difference time-domain (FDTD) method by discretizing space and time into a regular grid, have been developed for calculation of absorbed energy by organisms from nonionizing radiation exposures [48]. Commercial or open-source software packages such as MEEP [49], CST [50], and XFDTD [51] based on these computational methods are available and are used for bio-electromagnetics research using computational animal models.

3.5 Applications of computational mouse models

3.5.1 Ionizing radiation dosimetry

Computational animal models were commonly used to estimate absorbed dose distribution and the dose deposited within organs of the animal for ionizing radiation exposures from external radiation sources or internal radiotracers. The basic measures of absorbed fraction (AF), specific absorbed fraction (SAF) and *S*-values are defined in the medical internal radiation dose (MIRD) standard for evaluating internal radiation dose. AF represents the fraction of absorbed energy in the target organ released from a source organ, SAF describes the absorbed fraction in the target organ per unit mass of the target organ, and *S*-value is the absorbed dose in the target organ per unit activity in the source organ. AF, SAF, and *S*-values are used for calculation of individual organ absorbed dose in mice. AF and SAF are commonly estimated for mono-energetic photons and electrons, and their values (ranging between 0 and 1) depend on the type and energy of the radiation, and the shape and mass of organ in the computational mouse model. *S*-values depend on the decay scheme of the radionuclide, the type, energy and yield of emitted radiation per nuclear transformation, and the shape and mass of organ in the computational mouse model. Various groups have already reported on datasets of AF or SAF and *S*-values in organs of mice for mono-energetic photons and electrons and several radionuclides, including ^{18}F, ^{32}P, ^{64}Cu, ^{90}Y, ^{111}In, ^{124}I, ^{131}I, ^{149}Pm, ^{153}Sm, ^{166}Ho, ^{177}Lu and ^{188}Re [9, 13, 15, 16, 19, 22, 25–29, 52–59]. These data were used for calculation of organ absorbed dose and dose of radiopharmaceuticals in preclinical experiments [60–71] and in radiation dosimetry of environments in Mayak and Fukushima contaminated with radionuclides released from nuclear accidents [72, 73]. These calculated doses in the computational mouse models based on Monte Carlo simulations were applied to molecular radiotherapy and diagnostic nuclear medical imaging for investigating the patterns of uptake and biodistribution data of radiopharmaceuticals [74–85]. Mohammadi *et al* [23] investigated the effect of voxel size on SAF and *S*-value for Digimouse models with 0.1 and 0.4 mm voxel sizes, and significant differences were reported for cross-absorbed *S*-value or SAF for some organs. Tables 3.2 and 3.3 compare self-absorbed and cross-absorbed *S*-values for the Digimouse models with the different voxel sizes. Some parameters such as the shape, size, and mass of organs in computational mouse models have been reported as factors influencing mouse internal radiation dosimetry [11, 23, 27, 86]. The different mouse dosimetric results were reported for stylized and voxel-based models [87]; however, the discrepancy between AF of the two models for the whole-body mouse was determined to be less than 2%. Model resolution, organ segmentation, tissue density, and spatial sampling when using hybrid models for dose simulation were reported as the parameters with a significant effect on *S*-values [26]. Remarkable discrepancies of up to 160%, between the *S*-values of some organs were observed for the different mouse models, which was not directly correlated with mass variations. Several studies have been performed on the calculation of organ dose conversion coefficients for mice exposed to external ionizing radiation under ideal irradiation conditions [24], therapeutic irradiation conditions [21, 88, 89], diagnostic imaging x-ray irradiations [32, 35, 90], low-dose irradiation [91], analytical dose modeling for proton therapy [92], micro-beam radiation therapy [93, 94], and scattered dose under x-ray irradiation [95, 96]. Optimization of CT scanning protocols for PET/CT modalities and the effect of respiratory motion on dose delivery were reported for mice using the computational mouse models [97, 98].

Table 3.2. Self-absorbed *S*-values (Gy/Bq.s) in some organs of the Digimouse models with the voxel size of 0.1 and 0.4 mm. Source and target organs are the same. Reproduced with permission from Mohammadi and Kinase [23]. Copyright Clearance Center's RightsLink®service.

Nuclide	0.1 mm cube	0.4 mm cube	Diff. (%)
	Source = Eyes		
^{131}I	3.46×10^{-9}	3.63×10^{-9}	−4.93
^{153}Sm	4.92×10^{-9}	5.19×10^{-9}	−5.67
^{188}Re	4.40×10^{-9}	4.55×10^{-9}	−3.44
^{90}Y	3.89×10^{-9}	4.00×10^{-9}	−3.02
	Source = Cerebellum		
^{131}I	7.67×10^{-10}	7.33×10^{-10}	4.46
^{153}Sm	1.08×10^{-9}	1.03×10^{-9}	4.18
^{188}Re	1.32×10^{-9}	1.28×10^{-9}	3.34
^{90}Y	1.26×10^{-9}	1.22×10^{-9}	3.10
	Source = Kidneys		
^{131}I	5.53×10^{-11}	5.49×10^{-11}	0.79
^{153}Sm	7.77×10^{-11}	7.71×10^{-11}	0.67
^{188}Re	1.47×10^{-10}	1.46×10^{-10}	0.57
^{90}Y	1.53×10^{-10}	1.52×10^{-10}	0.54
	Source = Lungs		
^{131}I	1.61×10^{-10}	1.58×10^{-10}	1.79
^{153}Sm	2.30×10^{-10}	2.27×10^{-10}	1.48
^{188}Re	2.38×10^{-10}	2.36×10^{-10}	0.96
^{90}Y	2.21×10^{-10}	2.19×10^{-10}	0.77

Table 3.3. Cross-absorbed *S*-values (Gy/Bq.s) in some organs of the Digimouse models with the voxel size of 0.1 and 0.4 mm. Reproduced with permission from Mohammadi and Kinase [23]. Copyright Clearance Center's RightsLink®service.

Nuclide	0.1 mm cube	0.4 mm cube	Diff. %
	Source = Eyes, Target = Lachrymal		
^{131}I	4.53×10^{-11}	6.11×10^{-11}	−34.76
^{153}Sm	6.66×10^{-11}	8.56×10^{-11}	−28.54
^{188}Re	3.18×10^{-10}	3.43×10^{-10}	−7.68
^{90}Y	3.46×10^{-10}	3.69×10^{-10}	−6.64
	Source = Cerebellum, Target = Rest of brain		
^{131}I	1.54×10^{-11}	1.41×10^{-11}	8.39
^{153}Sm	2.07×10^{-11}	1.90×10^{-11}	8.23
^{188}Re	1.12×10^{-10}	1.09×10^{-10}	2.88
^{90}Y	1.30×10^{-10}	1.27×10^{-10}	2.35
	Source = Kidneys, Target = Pancreas		
^{131}I	3.44×10^{-12}	3.71×10^{-12}	−7.84
^{153}Sm	4.53×10^{-12}	4.83×10^{-12}	−6.62
^{188}Re	3.34×10^{-11}	3.36×10^{-11}	−0.53
^{90}Y	4.15×10^{-11}	4.16×10^{-11}	−0.37
	Source = Lungs, Target = Liver		
^{131}I	1.08×10^{-12}	1.14×10^{-12}	−5.13
^{153}Sm	1.34×10^{-12}	1.41×10^{-12}	−4.83
^{188}Re	8.78×10^{-12}	8.90×10^{-12}	−1.28
^{90}Y	1.12×10^{-11}	1.13×10^{-11}	−1.11

3.5.2 Nonionizing radiation dosimetry

For nonionizing radiation, computational mouse models were extensively used for evaluation of biological effects caused by heat produced by radiofrequency (RF)-emitting devices such as electric power lines, high-field MRI and wireless cellular phones. There is great interest in studying nonionizing radiation effects due to the increased use of nonionizing radiation-emitting devices. The localized specific absorption rate (SAR), defined as the mass normalized rate of energy absorbed by the body, is a fundamental metric for limiting human exposure to RF electromagnetic fields. SAR values were reported for mice under RF exposure of 900 MHz and 1.5 GHz cellular phone electromagnetic field [20, 99, 100] and for mice exposed to 2.45 GHz WiFi frequencies [101, 102]. A computational pregnant mouse model with eight fetuses was developed to support RF dosimetry studies [20]. The SAR values in the fetuses were found to be around 86% of that of the dam, whereas the peak temperature increase was around 45% of the value for the dam.

3.5.3 Medical imaging physics

Accurate quantification of the uptake and pharmacokinetics of radiotracers in small structures within rodents, especially the brain, requires the design and development of ultra-high-resolution scanner systems as well as novel image reconstruction techniques for improving image quality. Several multimodal preclinical imaging systems specifically designed for rodent research have been proposed by various research groups within the past decade. To evaluate the designs, and optimize the entire image acquisition, reconstruction and analysis process prior to building the first prototype scanners, computational mouse models are used in combination with simulation tools. Production of simulated imaging data from computational models with precisely known activity distributions and attenuation characteristics provides quantitative information regarding the effect of physical degradation factors such as scatter fraction on the imaging process [103–107], and allows a detailed comparative evaluation of the performance of alternative design concepts for medical imaging systems to be performed with precise knowledge of ground truth values [108–125].

Simulating image acquisition with different imaging modalities using realistic mouse models is also extremely valuable for researchers working to develop and evaluate new methods for image registration/segmentation [126–136], reconstruction [137–143], and quantitative analysis [144–151]. The MOBY computational mouse phantom has been used for each of these purposes, two examples of which will now be described. Li *et al* [137] used a MOBY phantom to generate a map of realistic [^{18}F]FDG uptake distribution to evaluate the performance of their proposed virtual cylinder method for system matrix compression in highly pixelated DOI (depth-of-interaction) PET scanners. Reilhac *et al* evaluated the impact of single-mouse versus simultaneous dual-mouse imaging on the ability of a PET scanner to detect biological differences in control versus pathological groups in simulated whole-body and brain studies [152]. A validated Monte Carlo simulation model of a Siemens Inveon preclinical PET scanner [113] was used together with the Digimouse (brain) and MOBY (whole-body) computational models to simulate control and pathological animals, and the degradation in detection of the known biological variations were quantified. Figure 3.5 shows a dual-mode simulation configuration, where two Moby phantoms have been simulated side-by-side inside a simulated Siemens Inveon Focus F220 preclinical PET/CT scanner.

The Digimouse model was also used as a heterogeneous medium for simulation of light propagation to assess the effect of various parameters on optical molecular imaging techniques and to improve efficiency and accuracy of simulation [153–160]. In the same way, the performance of novel image reconstruction techniques in bioluminescence tomography (BLT) imaging systems were explored using the Digimouse and MOBY models [161–171]. For fluorescence molecular tomography (FMT), several simulation studies with both mouse models were also performed to confirm the improvement of the performance of the systems [172–182].

3.6 Summary

The mouse computational models developed gradually from stylized models based on simple shape modeling of the internal organs and the exterior body in 1994, to voxel-based models with increasingly realistic anatomy extracted from medical imaging data of mice, and more recently, hybrid computational models which

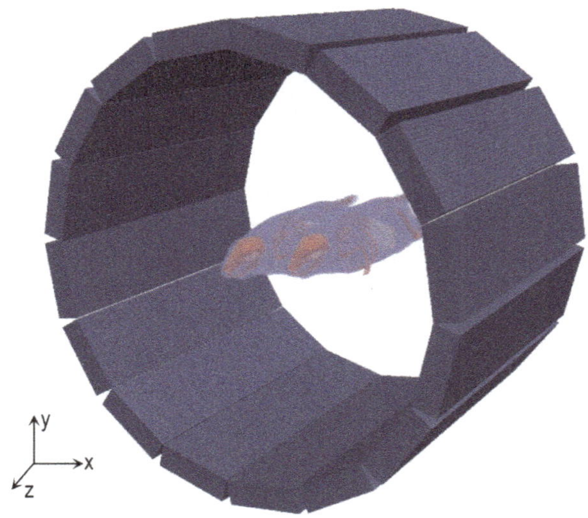

Figure 3.5. A dual-mode simulation configuration of two Moby phantoms inside a Siemens Inveon Focus F220 preclinical PET/CT scanner.

combine the realism of voxel-based models with the simple shape modeling of stylized models. The computational models have become progressively more sophisticated over the past decade, with ever-more realistic anatomical models becoming available. Around 26 computational mouse models have been constructed for various research projects in ionizing and nonionizing radiation dosimetry, preclinical imaging including multimodality imaging and instrumentation, image processing and analysis.

The three models of stylized, voxel and hybrid models differ in the aspects of anatomical realism, geometrical flexibility, and compatibility with Monte Carlo simulation tools. The stylized models have unrealistic anatomy which results in high uncertainty and significant deviation of dosimetric results under varied conditions. Unrealistic anatomy is not a problem for hybrid and voxel models; however, voxel size was reported as an important parameter which can affect radiation dosimetry due to the cube-shaped surface of voxels [23]. The stylized and hybrid models have smooth organ surfaces and enable easy deformation of organ shape during respiratory motion and physiological cardiac motion. The stylized and voxel models have the simplest geometries for easy integration with most Monte Carlo simulation tools. However, the majority of hybrid models cannot be directly integrated in popular Monte Carlo simulation packages, and thus it is necessary to convert hybrid models to equivalent voxel models using voxelizing tools. Recently, there have been some efforts to insert polygon mesh and NURBS surfaces to the geometry definition of some Monte Carlo simulation packages, for instance, the new version of GEANT4 package can directly implement polygon mesh surfaces for simulations [183] and results have been reported for computational human models [184, 185].

One important issue that should be considered is the usage of convenient mouse computational models for preclinical research. Significant individual variations of dosimetric results have been reported for different mouse models of the same mouse with close weights [25, 26]. Therefore, it is necessary to choose the most convenient computational mouse models based on anatomical geometry of the experimental mouse for radiation dosimetry.

The approach of using computational mouse models together with a range of simulation tools is very useful for improving the efficiency of experimental approach using physical phantoms or real mice, due to high cost and complexity of organizing experimental research. The computational models can be used to generate data under controlled conditions, by defining the desired anatomical and physiological functions as reference data for evaluation of performance of imaging devices, accuracy of image reconstruction techniques, or estimation of radiation dose. Furthermore, a large population of the computational mouse models may be generated for medical physics research because of the simplicity of creating various anatomies and physiological conditions.

Choosing the most convenient computational model for a specific application is a fundamental challenge in this context. For radiation dosimetry calculations, self-absorbed *S*-values and cross-absorbed *S*-values strongly depend on the organ mass of the model and the distance between the organs. Therefore, a mouse model having the same body weight, length, age, and strain, as a real experimental mouse would be the most convenient choice. Stylized mouse models would be suitable for environmental radiation protection where an average model is sufficient and the accuracy of dosimetric calculated data is limited by anatomic variations between individual mice. Hybrid mouse models are the most convenient models in molecular imaging applications involving dynamic studies considering organ motions due to respiration and heart beating.

Computational mouse models play an important role in the validation of new imaging techniques and reducing the number of experiments that need to be performed with laboratory mice for molecular imaging and targeted molecular therapy.

For preclinical studies mice are the most important laboratory animal, and thus the development of dosimetric datasets for mice of different strains, weights, and ages could be valuable for assessing radiation dose from multimodality imaging systems. Meanwhile, with the development of small animal conformal radiation therapy system [186], computational mouse models would be necessary for translation of preclinical data for application in human patients.

References

[1] Xie T and Zaidi H 2016 Development of computational small animal models and their applications in preclinical imaging and therapy research *Med. Phys.* **43** 111–31

[2] Zaidi H and Tsui B M W 2009 Review of computational anthropomorphic anatomical and physiological models *Proc. IEEE* **97** 1938–53

[3] Hui T E, Fisher D R, Kuhn J A, Williams L E, Nourigat C, Badger C C, Beatty B G and Beatty J D 1994 A mouse model for calculating cross-organ beta doses from yttrium-90-labeled immunoconjugates *Cancer* **73** 951–7

[4] Yoriyaz H and Stabin M 1997 Electron and photon transport in a model of a 30 g mouse [abstract] *J. Nucl. Med.* **38** (suppl) 228 P

[5] Kennel S J, Stabin M, Yoriyaz H, Brechbiel M and Mirzadeh S 1999 Treatment of lung tumor colonies with ^{90}Y targeted to blood vessels: Comparison with the α-particle emitter213Bi *Nucl. Med. Biol.* **26** 149–57

[6] Dhenain M, Ruffins S W and Jacobs R E 2001 Three-dimensional digital mouse atlas using high-resolution MRI *Dev. Biol.* **232** 458–70

[7] Flynn A A, Green A J, Pedley R B, Boxer G M, Boden R and Begent R H 2001 A mouse model for calculating the absorbed beta-particle dose from ^{131}I- and ^{90}Y-labeled immunoconjugates, including a method for dealing with heterogeneity in kidney and tumor *Radiat. Res.* **156** 28–35

[8] Wang J and Fujiwara O 2001 A novel setup for small animal exposure to near fields to test biological effects of cellular telephones *IEICE Trans. Commun. E84–B* 3050–9

[9] Kolbert K S, Watson T, Matei C, Xu S, a Koutcher J and Sgouros G 2003 Murine S factors for liver, spleen, and kidney *J. Nucl. Med.* **44** 784–91

[10] Funk T, Sun M and Hasegawa B H 2004 Radiation dose estimate in small animal SPECT and PET *Med. Phys.* **31** 2680–6

[11] Hindorf C, Ljungberg M and Strand S-E 2004 Evaluation of parameters influencing S values in mouse dosimetry *J. Nucl. Med.* **45** 1960–5

[12] Segars W P, Tsui B M W, Frey E C, Johnson G A and Berr S S 2004 Development of a 4-D digital mouse phantom for molecular imaging research *Mol. Imaging Biol.* **6** 149–59

[13] Miller W H *et al* 2005 Evaluation of beta-absorbed fractions in a mouse model for ^{90}Y, ^{188}Re, ^{166}Ho, ^{149}Pm, ^{64}Cu, and ^{177}Lu Radionuclides *Cancer Biother. Radiopharm.* **20** 436–49

[14] Sato Y 2008 Internal dose distribution of ^{90}Y beta-ray source implanted in a small phantom simulating a mouse *Radioisotopes* **57** 385–91

[15] Stabin M G, Peterson T E, Holburn G E and Emmons M A 2006 Voxel-based mouse and rat models for internal dose calculations *J. Nucl. Med.* **47** 655–9

[16] Bitar A, Lisbona A, Thedrez P, Sai Maurel C, Le Forestier D, Barbet J and Bardies M 2007 A voxel-based mouse for internal dose calculations using Monte Carlo simulations (MCNP) *Phys. Med. Biol.* **52** 1013–25

[17] Dogdas B, Stout D, Chatziioannou A F and Leahy R M 2007 Digimouse: A 3D whole body mouse atlas from CT and cryosection data *Phys. Med. Biol.* **52** 577–87

[18] Zhang X, Tian J, Feng J, Zhu S and Yan G 2009 An anatomical mouse model for multimodal molecular imaging *Proc. 31st Annu. Int. Conf. IEEE Eng. Med. Biol. Soc. Eng. Futur. Biomed. EMBC 2009* 5817–20

[19] Keenan M A, Stabin M G, Segars W P and Fernald M J 2010 RADAR realistic animal model series for dose assessment *J. Nucl. Med.* **51** 471–6

[20] McIntosh R L, Deppeler L, Oliva M, Parente J, Tambuwala F, Turner S, Winship D and Wood A W 2010 Comparison of radiofrequency exposure of a mouse dam and foetuses at 900 MHz *Phys. Med. Biol.* **55** N111–22

[21] Chow J C L, Leung M K K, Lindsay P E and Jaffray D A 2010 Dosimetric variation due to the photon beam energy in the small-animal irradiation: A Monte Carlo study *Med. Phys.* **37** 5322–9

[22] Larsson E, Ljungberg M, Strand S E and Jönsson B A 2011 Monte Carlo calculations of absorbed doses in tumours using a modified MOBY mouse phantom for pre-clinical dosimetry studies *Acta Oncol. (Madr).* **50** 973–80

[23] Mohammadi A and Kinase S 2011 Influence of voxel size on specific absorbed fractions and S-values in a mouse voxel phantom *Radiat. Prot. Dosim.* **143** 258–63
[24] Zhang X, Xie X, Cheng J, Ning J, Yuan Y, Pan J and Yang G 2012 Organ dose conversion coefficients based on a voxel mouse model and MCNP code for external photon irradiation *Radiat. Prot. Dosim.* **148** 9–19
[25] Xie T and Zaidi H 2013 Monte Carlo-based evaluation of S-values in mouse models for positron-emitting radionuclides *Phys. Med. Biol.* **58** 169–82
[26] Mauxion T, Barbet J, Suhard J, Pouget J P, Poirot M and Bardiès M 2013 Improved realism of hybrid mouse models may not be sufficient to generate reference dosimetric data *Med. Phys.* **40** 052501-11
[27] Kostou T, Papadimitroulas P, Loudos G and Kagadis G C 2016 A preclinical simulated dataset of S-values and investigation of the impact of rescaled organ masses using the MOBY phantom *Phys. Med. Biol.* **61** 2333–55
[28] Mendes B M, de Almeida I G, Trindade B M, Fonseca T C F and de Campos T P R 2017 Development of a mouse computational model for MCNPx based on Digimouse® images and dosimetric assays, Brazilian *J. Pharm. Sci.* **53** 1–12
[29] Locatelli M, Miloudi H, Autret G, Balvay D, Desbrée A, Blanchardon E and Bertho J 2017 RODES software for dose assessment of rats and mice contaminated with radionuclides *J. Radiol. Prot.* **37** 214–29
[30] Zaidi H and El Naqa I 2010 PET-guided delineation of radiation therapy treatment volumes: A survey of image segmentation techniques *Eur. J. Nucl. Med. Mol. Imaging* **37** 2165–87
[31] Ahrens J, Law C, Geveci B, Ahrens J and Geveci C L B 2005 Paraview: An end user tool for large data visualization *Energy* **836** 717–31
[32] Boone J M, Velazquez O and Cherry S R 2004 *Small-animal x-ray dose from micro-CT Mol. Imaging* **3** 149–58
[33] Badea C T, Drangova M, Holdsworth D W and Johnson G A 2008 *In vivo* small-animal imaging using micro-CT and digital subtraction angiography *Phys. Med. Biol.* **53** R319–350
[34] Ford N L, Thornton M M and Holdsworth D W 2003 Fundamental image quality limits for microcomputed tomography in small animals *Med. Phys.* **30** 2869–77
[35] Taschereau R, Chow P L and Chatziioannou A F 2006 Monte Carlo simulations of dose from microCT imaging procedures in a realistic mouse phantom *Med. Phys.* **33** 216–24
[36] Carlson S K, Classic K L, Bender C E and Russell S J 2007 Small animal absorbed radiation dose from serial micro-computed tomography imaging *Mol. Imaging Biol.* **9** 78–82
[37] Badea C T, Fubara B, Hedlund L W and Johnson G A 2005 4-D micro-CT of the mouse heart *Mol. Imaging* **4** 110–16
[38] Joshi A, Chaudhari A J and Li C *et al* 2009 Posture matching and elastic registration of a mouse atlas to surface topography range data *2009 IEEE Int. Symp. Biomed. Imaging From Nano to Macro* 366–9
[39] Wang H, Stout D B and Chatziioannou A F 2014 A deformable atlas of the laboratory mouse *Mol. Imaging Biol.* **17** 18–28
[40] Zaidi H 1999 Relevance of accurate Monte Carlo modeling in nuclear medical imaging *Med. Phys.* **26** 574–608
[41] Rogers D W O 2006 Fifty years of Monte Carlo simulations for medical physics *Phys. Med. Biol.* **51** R287–301

[42] Kawrakow I and Rogers D W O 2017 The EGSnrc code system: Monte Carlo simulation of electron and photon transport *NRC Institute for National Measurement Standards; National Research Council Canada Technical Report PIRS-701* 2001–15

[43] Briesmeister J F 2000 MCNPTM—A general Monte Carlo N-particle transport code *Report LA-13709-M* (Los Alamos National Laboratory, Los Alamos, NM)

[44] Agostinelli S, Allison J, Amako K, Apostolakis J and Araujo H *et al* 2003 GEANT4 - A simulation toolkit *Nucl. Instrum. Meth. Phys. Res. Sect.* A **506** 250–303

[45] GATE: simulations of preclinical and clinical scans in emission tomography, transmission tomography and radiation therapy. http://www.opengatecollaboration.org/ (accessed March 7, 2018).

[46] Zhu C and Liu Q 2013 Review of Monte Carlo modeling of light transport in tissues *J. Biomed. Opt.* **18** 50902

[47] Li H, Tian J, Zhu F, Cong W, Wang L V, Hoffman E A and Wang G 2004 A mouse optical simulation environment (MOSE) to investigate bioluminescent phenomena in the living mouse with the Monte Carlo method *Acad. Radiol.* **11** 1029–38

[48] Sheng X-Q and Song W 2012 *Essentials of Computational Electromagnetics* (Singapore: Wiley)

[49] Oskooi A F, Roundy D, Ibanescu M, Bermel P, Joannopoulos J D and Johnson S G 2010 Meep: A flexible free-software package for electromagnetic simulations by the FDTD method *Comput. Phys. Commun.* **181** 687–702

[50] CST, computer simulation technology, https://www.cst.com/ (accessed March 2, 2018)

[51] REMCOM, https://www.remcom.com/ (accessed March 8, 2018)

[52] Bitar A, Lisbona A and Bardiès M 2007 S-factor calculations for mouse models using Monte-Carlo simulations *Q. J. Nucl. Med. Mol. Imaging* **51** 343–51

[53] Kinase S, Mohammadi A, Takahashi M, Saito K, Zankl M and Kramer R 2011 Computer simulations for internal dosimetry using voxel models *Radiat. Prot. Dosim.* **146** 191–4

[54] Larsson E, Strand S-E, Ljungberg M and Jönsson B-A 2007 Mouse S-factors based on Monte Carlo simulations in the anatomical realistic Moby phantom for internal dosimetry *Cancer Biother. Radiopharm.* **22** 438–42

[55] Mohammadi A and Kinase S 2011 Monte Carlo simulations of photon specific absorbed fractions in a mouse voxel *Prog. Nucl. Sci. Technol.* **1** 126–9

[56] Mohammadi A and Kinase S 2011 Electron absorbed fractions and s values in a voxel-based mouse phantom *Radioisotopes* **60** 505–12

[57] Sinha A, Patni H K, Dixit B M, Painuly N K and Singh N 2016 Estimation of photon specific absorbed fractions in digimouse voxel phantom using Monte Carlo simulation code FLUKA *J. Biomed. Phys. Eng.* **6** 209–16

[58] Sinha A, Singh N, Dixit B M and Painuly K N 2018 Assessment of S values for I-131 and Y-90 based on the Digimouse voxel phantom *J. Cancer Res. Ther.* 1–5

[59] Woo S, Huang S, Kang J H, Lim S M and Seo Y 2015 Object-specific internal dosimety studies of organs and tumors in a murine model of neuroblastoma using Monte Carlo simulation and voxelized mouse phantom *J. Nucl. Med.* **56** 1286

[60] Luna-Gutiérrez M, Ferro-Flores G, Ocampo-García B, Jiménez-Mancilla N, Morales-Avila E, De Leõn-Rodríguez L and Isaac-Olivé K 2012 ^{177}Lu-labeled monomeric, dimeric and multimeric RGD peptides for the therapy of tumors expressing $\alpha(\nu)\beta(3)$ integrins *J. Label. Compd. Radiopharm.* **55** 140–8

[61] Bertho J-M, Synhaeve N, Miloudi H, Stefani J, Desbrée A, Blanchardon E and Dublineau I 2012 Absorbed radiation doses due to chronic ingestion of cesium-137 or strontium-90 by mice *Radioprotection* **47** 219–30
[62] Song H *et al* 2008 ^{213}Bi (α-emitter)-antibody targeting of breast cancer metastases in the neu-N transgenic mouse model *Cancer Res.* **68** 3873–80
[63] Meléndez-Alafort L, Nadali A, Zangoni E, Banzato A, Rondina M, Rosato A and Mazzi U 2009 Biokinetic and dosimetric studies of ^{188}Re-hyaluronic acid: A new radiopharmaceutical for treatment of hepatocellular carcinoma *Nucl. Med. Biol.* **36** 693–701
[64] Rodríguez-Cortés J, Arteaga De Murphy C, Ferro-Flores G, Pedraza-López M and Murphy-Stack E 2007 Biokinetics and dosimetry with ^{177}Lu-DOTA-TATE in athymic mice with induced pancreatic malignant tumours: Preclinical studies *Radiat. Eff. Defects Solids* **162** 791–6
[65] Balkin E R, Jia F, Miller W H and Lewis M R 2011 *In vitro* evaluation of targeted antisense ^{177}Lu radiotherapy *Anticancer Res.* **31** 3143–9
[66] Fischer E, Grünberg J, Cohrs S, Hohn A, Waldner-Knogler K, Jeger S, Zimmermann K, Novak-Hofer I and Schibli R 2012 L1-CAM-targeted antibody therapy and ^{177}Lu-radioimmunotherapy of disseminated ovarian cancer *Int. J. Cancer.* **130** 2715–21
[67] Muthuswamy M S, Roberson P L and Buchsbaum D J 1998 A mouse bone marrow dosimetry model *J. Nucl. Med.* **39** 1243–7
[68] Green D J *et al* 2014 A preclinical model of CD38-pretargeted radioimmunotherapy for plasma cell malignancies *Cancer Res.* **74** 1179–89
[69] Frampas E, Maurel C, Le Saëc P R, Mauxion T, Faivre-Chauvet A, Davodeau F, Goldenberg D M, Bardiès M and Barbet J 2011 Pretargeted radioimmunotherapy of colorectal cancer metastases: Models and pharmacokinetics predict influence of the physical and radiochemical properties of the radionuclide *Eur. J. Nucl. Med. Mol. Imaging* **38** 2153–64
[70] Lazewatsky J, Ding Y, Onthank D, Silva P, Solon E and Robinson S 2003 Radiation Dose to Abdominal Organs of the Mouse due to 90 Y in the Urinary Bladder *Cancer Biother. Radiopharm.* **18** 413–9
[71] Liu G, Dou S, Mardirossian G, He J, Zhang S, Liu X, Rusckowski M and Hnatowich D J 2006 Successful radiotherapy of tumor in pretargeted mice by 188Re-radiolabeled phosphorodiamidate morpholino oligomer, a synthetic DNA analogue *Clin. Cancer Res.* **12** 4958–64
[72] Malinovsky G P, Yarmoshenko I V, Starichenko V I and Lyubashevsky N M 2014 Assessment of radiation exposure of murine rodents at the EURT territories *Cent. Eur. J. Biol.* **9** 960–6
[73] Caffrey E A, Johansen M P and Higley K A 2015 Organ dose rate calculations for small mammals at maralinga, the nevada test site, hanford, and fukushima: A comparison of ellipsoidal and voxelized dosimetric methodologies *Radiat. Res.* **184** 433–41
[74] Balkin E R, Liu D, Jia F, Ruthengael V C, Shaffer S M, Miller W H and Lewis M R 2014 Comparative biodistributions and dosimetry of [^{177}Lu]DOTA-anti-bcl-2-PNA-Tyr3-octreotate and [^{177}Lu]DOTA-Tyr3-octreotate in a mouse model of B-cell lymphoma/leukemia *Nucl. Med. Biol.* **41** 36–42
[75] Bryan J N, Jia F, Mohsin H, Sivaguru G, Miller W H, Anderson C J, Henry C J and Lewis M R 2005 Comparative uptakes and biodistributions of internalizing vs. noninternalizing copper-64 radioimmunoconjugates in cell and animal models of colon cancer *Nucl. Med. Biol.* **32** 851–8

[76] Cambien B *et al* 2014 $^{99m}TcO_4^-$-, auger-mediated thyroid stunning: Dosimetric requirements and associated molecular events *PLoS One* **9** e92729
[77] Goddu S M, Bishayee A, Bouchet L G, Bolch W E, Rao D V and Howell R W 2000 Marrow toxicity of ^{33}P-versus ^{32}P-orthophosphate: Implications for therapy of bone pain and bone metastases *J. Nucl. Med.* **41** 941–51
[78] Kinase S, Matsuhashi S and Saito K 2009 Interspecies scaling of self-organ doses from a voxel mouse to voxel humans *Nucl. Technol.* **168** 154–7
[79] Melis M, Vegt E, Konijnenberg M W, De Visser M and Bijst M 2010 Nephrotoxicity in mice after repeated imaging using ^{111}In-Labeled peptides *J. Nucl. Med.* **51** 973–7
[80] Mohsin H *et al* 2011 Comparison of pretargeted and conventional CC49 radioimmunotherapy using^{149}Pm, ^{166}Ho, and ^{177}Lu *Bioconjug. Chem.* **22** 2444–52
[81] Taschereau R and Chatziioannou A F 2007 Monte Carlo simulations of absorbed dose in a mouse phantom from 18-fluorine compounds *Med. Phys.* **34** 1026–36
[82] Thompson S, Ballard B, Jiang Z, Revskaya E, Sisay N, Miller W H, Cutler C S, Dadachova E and Francesconi L C 2014 ^{166}Ho and ^{90}Y labeled 6D2 monoclonal antibody for targeted radiotherapy of melanoma: Comparison with188Re radiolabel *Nucl. Med. Biol.* **41** 276–81
[83] Ugur O, Kostakoglu L, Hui E T, Fisher D R, Garmestani K, Gansow O A, Cheung N K V and Larson S M 1996 Comparison of the targeting characteristics of various radioimmunoconjugates for radioimmunotherapy of neuroblastoma: Dosimetry calculations incorporating cross-organ beta doses *Nucl. Med. Biol.* **23** 1–8
[84] Ugur O *et al* 1995 Calculated and TLD-based absorbed dose estimates for I-131-labeled 3F8 monoclonal antibody in a human neuroblastoma xenograft nude mouse model *Nucl. Med. Biol.* **22** 87–93
[85] Aranda-Lara L, Ferro-Flores G, Azorín-Vega E, de F, Ramírez M, Jiménez-Mancilla N, Ocampo-García B, Santos-Cuevas C and Isaac-Olivé K 2016 Synthesis and evaluation of Lys1(α,γ-Folate)Lys3(177Lu-DOTA)-Bombesin(1–14) as a potential theranostic radiopharmaceutical for breast cancer *Appl. Radiat. Isot.* **107** 214–9
[86] Boutaleb S, Pouget J P, Hindorf C, Pèlegrin A, Barbet J, Kotzki P O and Bardiès M 2009 Impact of mouse model on preclinical dosimetry in targeted radionuclide therapy *Proc. IEEE* **97** 2076–85
[87] Mohammadi A, Kinase S and Saito K 2011 Comparison of photon and electron absorbed fractions in voxel-based and simplified phantoms for small animals *Prog. Nucl. Sci. Technol.* **2** 365–8
[88] Verhaegen F, Granton P and Tryggestad E 2011 Small animal radiotherapy research platforms *Phys. Med. Biol.* **56**
[89] van Hoof S J, Granton P V and Verhaegen F 2013 Development and validation of a treatment planning system for small animal radiotherapy: SmART-Plan *Radiother. Oncol.* **109** 361–6
[90] Lee C-L, Park S-J, Jeon P-H, Jo B-D and Kim H-J 2016 Dosimetry in small-animal CT using Monte Carlo simulations *J. Instrum.* **11** T01003
[91] Noh S J *et al* 2016 Evaluation of *in vivo* low-dose mouse irradiation system *J. Instrum.* **11** P03031
[92] Vanstalle M, Constanzo J, Karakaya Y, Finck C, Rousseau M and Brasse D 2018 Analytical dose modeling for preclinical proton irradiation of millimetric targets *Med. Phys.* **45** 470–8

[93] Lin H, Jing J, Xu L and Mao X 2017 Monte Carlo study of the influence of energy spectra, mesh size, high Z element on dose and PVDR based on 1-D and 3-D heterogeneous mouse head phantom for Microbeam Radiation Therapy *Phys. Medica.* **44** 96–107
[94] Potez M, Bouchet A, Wagner J, Donzelli M, Bräuer-Krisch E, Hopewell J W, Laissue J and Djonov V 2018 Effects of synchrotron x-ray micro-beam irradiation on normal mouse ear pinnae *Int. J. Radiat. Oncol.*
[95] Welch D, Turner L, Speiser M, Randers-Pehrson G and Brenner D J 2017 Scattered dose calculations and measurements in a life-like mouse phantom *Radiat. Res.* **187** 433–42
[96] Choi M, Ghammraoui B and Badano A 2017 Small-angle x-ray scattering characteristics of mouse brain: Planar imaging measurements and tomographic imaging simulations *PLoS One* **12** 1–14
[97] Yang C-C 2016 Optimization of CT scanning protocols for small animal PET/CT studies: A Monte Carlo simulation study *J. Med. Imaging Heal. Informatics* **6** 925–31
[98] van der Heyden B, van Hoof S J, Schyns L E J R and Verhaegen F 2017 The influence of respiratory motion on dose delivery in a mouse lung tumour irradiation using the 4D MOBY phantom *Br. J. Radiol.* **90** 20160419
[99] Wang J, Fujiwara O and Nojima T 2003 Realizing highly localized exposure in small animals with absorbing material covered holder to test biological effects of 1. 5 GHz cellular telephones *IEICE Trans. Commun. E86–5* 1660–5
[100] Wang J, Saito T and Fujiwara O 2004 Uncertainty evaluation of dosimetry due to plastic holder for restraining small animal in *in vivo* near field exposure setup *IEEE Trans. Electromagn. Compat.* **46** 263–7
[101] Pinto R, Lopresto V, Galloni P, Marino C, Mancini S, Lodato R, Pioli C and Lovisolo G A 2010 Dosimetry of a set-up for the exposure of newborn mice to 2.45-GHZ WiFi frequencies *Radiat. Prot. Dosim.* **140** 326–32
[102] Wang J, Fujiwara O and Ono T 2002 Dosimetry evaluation of a whole body exposure setup for small animal at 2.45 GHz *IEICE Trans. Commun. E85–B* 2963–5
[103] Könik A, Madsen M T and Sunderland J J 2012 GATE simulations of small animal SPECT for determination of scatter fraction as a function of object size *IEEE Trans. Nucl. Sci.* **59** 1887–91
[104] Konik A, Koesters T, Madsen M T and Sunderland J J 2011 Evaluation of attenuation and scatter correction requirements as a function of object size in small animal PET imaging *IEEE Trans. Nucl. Sci.* **58** 2308–14
[105] Konik A, Madsen M and Sunderland J 2009 GATE simulations of human and small animal pet for determination of scatter fractions as a function of object size 2009 *IEEE Nuclear Science Symp. Conf. Record (NSS/MIC)* **59** 1–4
[106] Prasad R and Zaidi H 2014 Scatter characterization and correction for simultaneous multiple small-animal PET imaging *Mol. Imaging Biol.* **16** 199–209
[107] Prasad R and Zaidi H 2012 A cone-shaped phantom for assessment of small animal pet scatter fraction and count rate performance *Mol. Imaging Biol.* **14** 561–71
[108] Sisniega A, Zbijewski W, Badal A, Kyprianou I S, Stayman J W, Vaquero J J and Siewerdsen J H 2013 Monte Carlo study of the effects of system geometry and antiscatter grids on cone-beam CT scatter distributions *Med. Phys.* **40**
[109] Brard E and Brasse D 2013 Axially oriented crystal geometry applied to small-animal PET system: A proof of concept *IEEE Trans. Nucl. Sci.* **60** 1471–7
[110] Funk T, Després P, Barber W C, Shah K S and Hasegawa B H 2006 A multipinhole small animal SPECT system with submillimeter spatial resolution *Med. Phys.* **33** 1259–68

[111] Huang S, Truong D, Wu H-M, Chatziioannou A F, Shao W, Wu A M and Phelps M E 2005 An internet-based 'kinetic imaging system' (KIS) for microPET *Mol. Imaging Biol.* **7** 330–41

[112] Van Holen R, Vandeghinste B, Deprez K and Vandenberghe S 2013 Design and performance of a compact and stationary microSPECT system *Med. Phys.* **40** 1–11

[113] Boisson F, Wimberley C J, Lehnert W, Zahra D, Pham T, Perkins G, Hamze H, Gregoire M C and Reilhac A 2013 NEMA NU 4–2008 validation and applications of the PET-SORTEO Monte Carlo simulations platform for the geometry of the Inveon PET preclinical scanner *Phys. Med. Biol.* **58** 6749–63

[114] Branco S, Jan S and Almeida P 2007 Monte Carlo simulations in small animal PET imaging *Nucl. Instrum. Meth. Phys. Res. Sect.* A **580** 1127–30

[115] Branderhorst W, Van Der Have F, Vastenhouw B, Viergever M A and Beekman F J 2012 Murine cardiac images obtained with focusing pinhole SPECT are barely influenced by extra-cardiac activity *Phys. Med. Biol.* **57** 717–32

[116] Chen C L, Wang Y, Lee J J S and Tsui B M W 2008 Integration of SimSET photon history generator in GATE for efficient Monte Carlo simulations of pinhole SPECT *Med. Phys.* **35** 3278–84

[117] Goorden M C and Beekman F J 2010 High-resolution tomography of positron emitters with clustered pinhole SPECT *Phys. Med. Biol.* **55** 1265–77

[118] Lartizien C, Kuntner C, Goertzen A L, Evans A C and Reilhac A 2007 Validation of PET-SORTEO Monte Carlo simulations for the geometries of the MicroPET R4 and Focus 220 PET scanners *Phys. Med. Biol.* **52** 4845–62

[119] Lee O and Ye J C 2013 Joint sparsity-driven non-iterative simultaneous reconstruction of absorption and scattering in diffuse optical tomography *Opt. Express* **21** 26589–604

[120] Mok G S P, Du Y, Wang Y, Frey E C and Tsui B M W 2010 Development and validation of a Monte Carlo simulation tool for multi-pinhole SPECT *Mol. Imaging Biol.* **12** 295–304

[121] Mok G S P, Wang Y and Tsui B M W 2009 Quantification of the multiplexing effects in multi-pinhole small animal SPECT: A simulation study *IEEE Trans. Nucl. Sci.* **56** 2636–43

[122] Mok G S P, Yu J, Du Y, Wang Y and Tsui B M W 2012 Evaluation of a multi-pinhole collimator for imaging small animals with different sizes *Mol. Imaging Biol.* **14** 60–9

[123] Shokouhi S, Wilson D W, Metzler S D and Peterson T E 2010 Evaluation of image reconstruction for mouse brain imaging with synthetic collimation from highly multiplexed SiliSPECT projections *Phys. Med. Biol.* **55** 5151–68

[124] Dartial N, Carlier T, Fusellier M, Desfontis J, Visvikis D, Bardies M and Hindorf C 2009 Monte Carlo simulation of the Inveon microPET system using GATE(2009) *J. Nucl. Med.* **50** 1417

[125] Li K, Safavi-Naeini M, Franklin D R, Petasecca M, Guatelli S, Rosenfeld A B, Hutton B F and Lerch M L F 2013 A feasibility study of PETiPIX: An ultra high resolution small animal PET scanner *J. Instrum.* **8** P12004

[126] Baiker M, Milles J, Dijkstra J, Henning T D, Weber A W, Que I, Kaijzel E L, Lwik C W G M, Reiber J H C and Lelieveldt B P F 2010 Atlas-based whole-body segmentation of mice from low-contrast Micro-CT data *Med. Image Anal.* **14** 723–37

[127] Cheng-Liao J and Qi J 2010 Segmentation of mouse dynamic PET images using a multiphase level set method *Phys. Med. Biol.* **55** 6549–69

[128] Clark D, Badea A, Liu Y, Johnson G A and Badea C T 2012 Registration-based segmentation of murine 4D cardiac micro-CT data using symmetric normalization *Phys. Med. Biol.* **57** 6125–45

[129] Khmelinskii A, Groen H C, Baiker M, de Jong M and Lelieveldt B P F 2012 Segmentation and visual analysis of whole-body mouse skeleton microSPECT *PLoS One* **7** e48976

[130] Maroy R, Boisgard R, Comtat C, Frouin V, Cathier P, Duchesnay E, Dollé F, Nielsen P E, Trébossen R and Tavitian B 2008 Segmentation of rodent whole-body dynamic PET images: An unsupervised method based on voxel dynamics *IEEE Trans. Med. Imaging* **27** 342–54

[131] Badea C T, Schreibmann E and Fox T 2008 A registration based approach for 4D cardiac micro-CT using combined prospective and retrospective gating *Med. Phys.* **35** 1170–9

[132] Gutierrez D F and Zaidi H 2012 Automated analysis of small animal PET studies through deformable registration to an atlas *Eur. J. Nucl. Med. Mol. Imaging* **39** 1807–20

[133] Moore S C, Southekal S, Park M A, McQuaid S J, Kijewski M F and Muller S P 2012 Improved regional activity quantitation in nuclear medicine using a new approach to correct for tissue partial volume and spillover effects *IEEE Trans. Med. Imaging* **31** 405–16

[134] Wang H, Stout D B and Chatziioannou A F 2012 Mouse atlas registration with non-tomographic imaging modalities—a pilot study based on simulation *Mol. Imaging Biol.* **14** 408–19

[135] Wang H, Stout D B and Chatziioannou A F 2012 Estimation of mouse organ locations through registration of a statistical mouse atlas with micro-CT images *IEEE Trans. Med. Imaging* **31** 88–102

[136] Wu H M, Kreissl M C, Schelbert H R, Ladno W, Prins M, Shoghi-Jadid K, Chatziioannou A, Phelps M E and Huang S C 2005 First-pass angiography in mice using FDG-PET: A simple method of deriving the cardiovascular transit time without the need of region-of-interest drawing *IEEE Trans. Nucl. Sci.* **52** 1311–5

[137] Li K, Safavi-Naeini M, Franklin D R, Han Z, Rosenfeld A B, Hutton B and Lerch M L F 2015 A new virtual ring-based system matrix generator for iterative image reconstruction in high resolution small volume PET systems *Phys. Med. Biol.* **60** 6949–73

[138] Reilhac A *et al* 2015 4D PET iterative deconvolution with spatiotemporal regularization for quantitative dynamic PET imaging *Neuroimage* **118** 484–93

[139] Cabello J and Rafecas M 2012 Comparison of basis functions for 3D PET reconstruction using a Monte Carlo system matrix *Phys. Med. Biol.* **57** 1759–77

[140] Cheng-Liao J and Qi J 2011 PET image reconstruction with anatomical edge guided level set prior *Phys. Med. Biol.* **56** 6899–918

[141] D'Ambrosio D, Fiacchi G, Marengo M, Boschi S, Fanti S and Spinelli A E 2010 Reconstruction of dynamic pet images using accurate system point spread function modeling: Effects on parametric images *J. Mech. Med. Biol.* **10** 73–94

[142] Quan E and Lalush D S 2009 A faster ordered-subset convex algorithm for iterative reconstruction in a rotation-free micro-CT system *Phys. Med. Biol.* **54** 1061–72

[143] Salomon A, Goldschmidt B, Botnar R, Kiessling F and Schulz V 2012 A self-normalization reconstruction technique for PET scans using the positron emission data *IEEE Trans. Med. Imaging* **31** 2234–40

[144] Dumouchel T and Dekemp R A 2011 Analytical-based partial volume recovery in mouse heart imaging *IEEE Trans. Nucl. Sci.* **58** 110–20

[145] Dumouchel T, Thorn S, Kordos M, Dasilva J, Beanlands R S B and Dekemp R A 2012 A three-dimensional model-based partial volume correction strategy for gated cardiac mouse PET imaging *Phys. Med. Biol.* **57** 4309–34
[146] Kesner A L, Dahlbom M, Huang S-C, Hsueh W-A, Pio B S, Czernin J, Kreissl M, Wu H-M and Silverman D H S 2006 Semiautomated analysis of small-animal PET data *J. Nucl. Med.* **47** 1181–6
[147] Khmelinskii A, Baiker M, Kaijzel E L, Chen J, Reiber J H C and Lelieveldt B P F 2011 Articulated whole-body atlases for small animal image analysis: Construction and applications *Mol. Imaging Biol.* **13** 898–910
[148] Maroy R, Boisgard R, Comtat C, Jego B, Fontyn Y, Jan S, Dubois A, Trébossen R and Tavitian B 2010 Quantitative organ time activity curve extraction from rodent PET images without anatomical prior *Med. Phys.* **37** 1507–17
[149] Zan Y, Boutchko R, Huang Q, Li B, Chen K and Gullberg G T 2013 Fast direct estimation of the blood input function and myocardial time activity curve from dynamic SPECT projections via reduction in spatial and temporal dimensions *Med. Phys.* **40** 1–10
[150] Su Y, Welch M J and Shoghi K I 2007 The application of maximum likelihood factor analysis (MLFA) with uniqueness constraints on dynamic cardiac microPET data *Phys. Med. Biol.* **52** 2313–34
[151] Su Y and Shoghi K I 2010 Single-input–dual-output modeling of image-based input function estimation *Mol. Imaging Biol.* **12** 286–94
[152] Reilhac A *et al* 2016 Simultaneous scanning of two mice in a small-animal PET scanner: A simulation-based assessment of the signal degradation *Phys. Med. Biol.* **61** 1371–88
[153] Yang D, Chen X, Ren S, Qu X, Tian J and Liang J 2013 Influence investigation of a void region on modeling light propagation in a heterogeneous medium *Appl. Opt.* **52** 400
[154] Shen H and Wang G 2011 A study on tetrahedron-based inhomogeneous Monte Carlo optical simulation *Biomed. Opt. Express.* **2** 44
[155] Shen H and Wang G 2010 A tetrahedron-based inhomogeneous Monte Carlo optical simulator *Phys. Med. Biol.* **55** 947–62
[156] Yang D, Chen X, Peng Z, Wang X, Ripoll J, Wang J and Liang J 2013 Light transport in turbid media with non-scattering, low-scattering and high absorption heterogeneities based on hybrid simplified spherical harmonics with radiosity model *Biomed. Opt. Express.* **4** 2209
[157] Li W, Yi H, Zhang Q, Chen D and Liang J 2012 Extended finite element method with simplified spherical harmonics approximation for the forward model of optical molecular imaging *Comput. Math. Methods Med.* **2012** 394374
[158] Lee O, Kim J M, Bresler Y and Ye J C 2011 Compressive diffuse optical tomography: Noniterative exact reconstruction using joint sparsity *IEEE Trans. Med. Imaging* **30** 1129–42
[159] Chen J and Intes X 2009 Time-gated perturbation Monte Carlo for whole body functional imaging in small animals *Opt. Express* **17** 19566
[160] Srinivasan S, Pogue B W, Davis S and Leblond F 2007 Improved quantification of fluorescence in 3-D in a realistic mouse phantom; Chance B, Alfano R R, Tromberg B J, Tamura M and Sevick-Muraca E M *Proc. SPIE* 64340S
[161] Dehghani H, Davis S C and Pogue B W 2008 Spectrally resolved bioluminescence tomography using the reciprocity approach *Med. Phys.* **35** 4863–71
[162] Wu Q, Feng J, Jia K and Wang X 2013 Improved reconstruction quality of bioluminescent images by combining SPequations and Bregman iteration method *Comput. Math. Methods Med.* **2013** 767296

[163] Naser M A, Patterson M S and Wong J W 2012 Self-calibrated algorithms for diffuse optical tomography and bioluminescence tomography using relative transmission images *Biomed. Opt. Express* **3** 2794
[164] He X, Liang J, Wang X, Yu J, Qu X, Wang X, Hou Y, Chen D, Liu F and Tian J 2010 Sparse reconstruction for quantitative bioluminescence tomography based on the incomplete variables truncated conjugate gradient method *Opt. Express* **18** 24825–41
[165] Yu J, Liu F, Wu J, Jiao L and He X 2010 Fast source reconstruction for bioluminescence tomography based on sparse regularization *IEEE Trans. Biomed. Eng.* **57** 2583–6
[166] Chen X, Yang D, Qu X, Hu H, Liang J, Gao X and Tian J 2012 Comparisons of hybrid radiosity-diffusion model and diffusion equation for bioluminescence tomography in cavity cancer detection *J. Biomed. Opt.* **17** 66015
[167] Naser M A, Patterson M S and Wong J W 2014 Algorithm for localized adaptive diffuse optical tomography and its application in bioluminescence tomography *Phys. Med. Biol.* **59** 2089–109
[168] Alexandrakis G, Rannou F R and Chatziioannou A F 2006 Effect of optical property estimation accuracy on tomographic bioluminescence imaging: Simulation of a combined optical-PET (OPET) system *Phys. Med. Biol.* **51** 2045–53
[169] Alexandrakis G, Rannou F R and Chatziioannou A F 2005 Tomographic bioluminescence imaging by use of a combined optical-PET (OPET) system: A computer simulation feasibility study *Phys. Med. Biol.* **50** 4225–41
[170] Yu J, He X, Geng G, Liu F and Jiao L C 2013 Hybrid multilevel sparse reconstruction for a whole domain bioluminescence tomography using adaptive finite element *Comput. Math. Methods Med.* **2013** 548491
[171] Lv Y, Tian J, Cong W, Wang G, Yang W, Qin C and Xu M 2007 Spectrally resolved bioluminescence tomography with adaptive finite element analysis: Methodology and simulation *Phys. Med. Biol.* **52** 4497–512
[172] Wang X, Zhang B, Cao X, Liu F, Luo J and Bai J 2013 Acceleration of early-photon fluorescence molecular tomography with graphics processing units *Comput. Math. Meth. Med.* **2013** 1–9
[173] Liu X, Zhang B, Luo J and Bai J 2012 Principal component analysis of dynamic fluorescence tomography in measurement space *Phys. Med. Biol.* **57** 2727–42
[174] Raymond S B, Kumar A T N, Boas D A and Bacskai B J 2009 Optimal parameters for near infrared fluorescence imaging of amyloid plaques in Alzheimer's disease mouse models *Phys. Med. Biol.* **54** 6201–16
[175] Correia T, Aguirre J, Sisniega A, Chamorro-Servent J, Abascal J, Vaquero J J, Desco M, Kolehmainen V and Arridge S 2011 Split operator method for fluorescence diffuse optical tomography using anisotropic diffusion regularisation with prior anatomical information *Biomed. Opt. Express* **2** 2632–48
[176] Liu X, Liu F and Bai J 2011 A linear correction for principal component analysis of dynamic fluorescence diffuse optical tomography images *IEEE Trans. Biomed. Eng.* **58** 1602–11
[177] Chen J, Venugopal V and Intes X 2011 Monte Carlo based method for fluorescence tomographic imaging with lifetime multiplexing using time gates *Biomed. Opt. Express* **2** 871
[178] Chaudhari A J, Ahn S, Levenson R, Badawi R D, Cherry S R and Leahy R M 2009 Excitation spectroscopy in multispectral optical fluorescence tomography: Methodology, feasibility and computer simulation studies *Phys. Med. Biol.* **54** 4687–704

[179] Dutta J, Ahn S, Joshi A A and Leahy R M 2010 Illumination pattern optimization for fluorescence tomography: Theory and simulation studies *Phys. Med. Biol.* **55** 2961–82
[180] Joshi A, Rasmussen J C, Sevick-Muraca E M, Wareing T A and McGhee J 2008 Radiative transport-based frequency-domain fluorescence tomography *Phys. Med. Biol.* **53** 2069–88
[181] Abascal J F P-J, Aguirre J, Chamorro-Servent J, Schweiger M, Arridge S, Ripoll J, Vaquero J J and Desco M 2012 Influence of absorption and scattering on the quantification of fluorescence diffuse optical tomography using normalized data *J. Biomed. Opt.* **17** 36013
[182] Zhang G, Liu F, Pu H, He W, Luo J and Bai J 2014 A direct method with structural priors for imaging pharmacokinetic parameters in dynamic fluorescence molecular tomography *IEEE Trans. Biomed. Eng.* **61** 986–90
[183] Han M C, Kim C H, Jeong J H, Yeom Y S, Kim S, Wilson P P H and Apostolakis J 2013 DagSolid: A new Geant4 solid class for fast simulation in polygon-mesh geometry *Phys. Med. Biol.* **58** 4595–609
[184] Kim C H, Jeong J H, Bolch W E, Cho K W and Hwang S B 2011 A polygon-surface reference Korean male phantom (PSRK-Man) and its direct implementation in Geant4 Monte Carlo simulation *Phys. Med. Biol.* **56** 3137–61
[185] Kim C H *et al* 2015 The reference phantoms: voxel vs polygon *Ann. ICRP* **45** 188–201
[186] Stojadinovic S *et al* 2007 MicroRT - Small animal conformal irradiator *Med. Phys.* **34** 4706–16

IOP Publishing

Computational Anatomical Animal Models
Methodological developments and research applications
Habib Zaidi

Chapter 4

Overview of computational rat models

Qian Liu, Guozhi Zhang, Tianwu Xie and Habib Zaidi

4.1 Introduction

Laboratory animals are commonly used in preclinical research as models of human disease to develop and test new treatments and to investigate the relationship between radiobiological effectiveness and radiation exposure. In this context, the rat is valued as a superb laboratory animal in biomedical research owing to the resemblances between the rodent genome and the human genome, and the relatively easy and lower cost breeding of laboratory rats compared to other animals [1]. It has therefore become a well-characterized animal model for investigating radiation-induced lung injury, biological response to radiation of different quality, and age-related effects of radiation exposure. The pros of rats for preclinical research are the large number of strains, the short gestation period and the economical and convenient housing. The cons are the different organ structure and physiology from humans, the accelerated life span and the small body weight. The Wistar rat, Sprague–Dawley (SD) rat and Fischer rat are the most popular strains adopted in preclinical research. SD rats were used to investigate the radiation countermeasures to mitigate radiation-induced lung fibrosis using a model of whole-thoracic irradiation [2] while Wistar rats have been used in studies of late-arising, non-stochastic type of injuries to several major organ systems after combined irradiations, e.g. chronic kidney injuries with mitigated responses of the central nervous system [3] and wound skin injury [4]. Computational models are used as surrogates to characterize the physics and anatomy of organism and are integrated in modeling tools to simulate the transport of microwaves, photons, electrons, and other sources of radiation in biological tissues, thus enabling the calculation of energy deposition from radiopharmaceuticals and evaluation of the impact of physical degrading factors of different medical imaging systems.

In this chapter, we will review existing computational rat models and in particular introduce the development and applications of computational rat models developed by our research groups.

doi:10.1088/2053-2563/aae1b4ch4

4.2 Overview of existing rat models

As of 2018, over 60 computational rat models have been reported in the literature. These computational models were distributed among three main rat strains: Sprague–Dawley rat, Wistar rat and Fischer rat. In 1997, Burkhardt *et al* [5] from the Swiss Federal Institute of Technology in Switzerland developed one of the earliest voxel-based rat models based on MR images and performed dosimetric analysis of rats in wireless communication systems with 900 MHz frequency. In the same year, Lapin *et al* [6] at the Northwestern University in USA developed a 350 g voxel-based SD rat model based on CT images and used it in non-ionizing dosimetry research. Mason *et al* [7] at the Brooks Air Force Base in USA constructed a 370 g rat model using MRI data and calculated the SARs in the rat body from implanted temperature probes, heating-sensitive paints and infrared imaging. Chou *et al* [8] at the City of Hope National Medical Center reported the first stylized SD rat model based on CT scans and calculated the SARs in this ellipsoidal rat model from handheld wireless telephones. In 2000, based on CT images, Watanabe *et al* [9] at the Ministry of Posts and Telecommunications in Japan developed three voxel-based Fischer models with body weights of 126 g, 263 g and 359 g, respectively, and estimated the SAR distributions in rat models from microwave exposure. Funk *et al* [10] from the University of California at San Francisco developed three stylized rat models of 200 g, 300 g and 400 g with fixed axes ratios according to the anatomical measurement and calculated the radiation dose from F-18, Tc-99m, Tl-201, In-111, I-123 and I-125. Konijnenberg *et al* [11] at the Tyco Healthcare in the Netherlands reported a stylized model of 386 g Wistar rat based on the anatomical measurement and calculated the *S*-values for Y-90, In-111 and Lu-177. Leveque *et al* [12] at the Centre National de la Recherche Scientifique in France developed a voxel-based rat model based on CT and MR images for calculating the RF dose of rats from 900 MHz GSM mobile phone fields. Schonborn *et al* [13] at the Swiss Federal Institute of Technology constructed voxel models of a 300 g male Wistar rat and a 370 g male SD rat for RF dosimetry exposing to 1.62 GHz microwave. Segars *et al* [14] at the Johns Hopkins University developed the first and the most widely used hybrid model of a male Wistar rat based on MR images. In 2006, Stabin *et al* [15] at the Vanderbilt University reported a 248 g SD rat model based on CT images and estimated the AF values for monoenergetic electrons and photons. Kainz *et al* [16] at the Center for Devices and Radiological Health in US Food and Drug Administration developed voxel models of four SD rats of 567 g, 479 g, 252 g and 228 g body weights based on MRI and cryo-sectional images and performed RF dosimetry studies in 902 and 1747 MHz microwave fields. Lopresto *et al* [17] at the ENEA in Italy constructed a stylized rat model to investigate the biological effects in rats following exposures to electromagnetic fields of GSM 1800 system. In 2008, the ICRP [18] announced a 314 g stylized rat model based on the anatomic data and provided dose conversion factors for small animals under external exposure to 75 radionuclides. Peixoto *et al* [19] from the Universidade Federal de Pernambuco in Brazil reported on a voxel-based 310 g Wistar rat model and calculated AFs for monoenergetic photons and electrons. Wu *et al* [20] at HUST in China developed a

156 g SD rat model based on cryo-sectional images and performed radiation dosimetry calculations for rats under external irradiation with photons. Wang *et al* [21] at the Nagoya Institute of Technology in Japan constructed a pregnant Fischer rat model based on MR images and calculated the RF doses for pregnant rats from 1.95 GHz cellular phones. Pain *et al* [22] at the Universités Paris 11/Paris 7 constructed a 284 g SD rat model for evaluating the uncertainties on radiotracer accumulation quantitation in beta microprobe studies. Zhang *et al* [23] at HUST in China developed a hybrid SD rat model using the boundary representation method for radiological imaging applications. Based on the MOBY rat model, Keenan *et al* [24] at Vanderbilt University generated five voxel models of Wistar rats of 200 g, 300 g, 400 g, 500 g, and 600 g body weights, respectively, and reported the radiation doses from monoenergetic electrons and photons as well as a number of radionuclides including F-18, P-32, I-124, Y-90, In-111, Lu-177. Xie *et al* [25] constructed a stylized SD rat model for calculating the radiation absorbed dose of monoenergetic electrons and photons. Arima *et al* [26] at the NICT in Japan developed 115 g, 314 g and 472 g rat models for RF dosimetry of rats in 1500 MHz microwave fields. Larsson *et al* [27] at Lund University in Sweden derived four Brown Norway rat models from the MOBY model and calculated *S*-values for Y-90 and Lu-177. Xie *et al* [28] at HUST constructed a 153 g SD rat model using cryo-sectional images for calculating the *S*-values and dose distributions of Y-90, I-131, Ho-166, and Re-188 in the liver. Xie *et al* [29] from Geneva University Hospital in Switzerland developed 10 voxel-based Wistar rat models at different ages and evaluated the effect of changing age on small-animal radiation dosimetry from internal exposure. In 2013, Xie *et al* [30] constructed seven rat models with different degrees of emaciation and obesity to assess the effect of emaciation and obesity on small-animal radiation dosimetry. Locatelli *et al* [31] at the Université Paris Descartes-Sorbonne Paris Cité generated voxel models of 35 g, 208 g and 537 g female and 38 g, 277 g and 530 g male C57BL/6 and reported the RODES software for dose assessment of laboratory rodents contaminated with radionuclides.

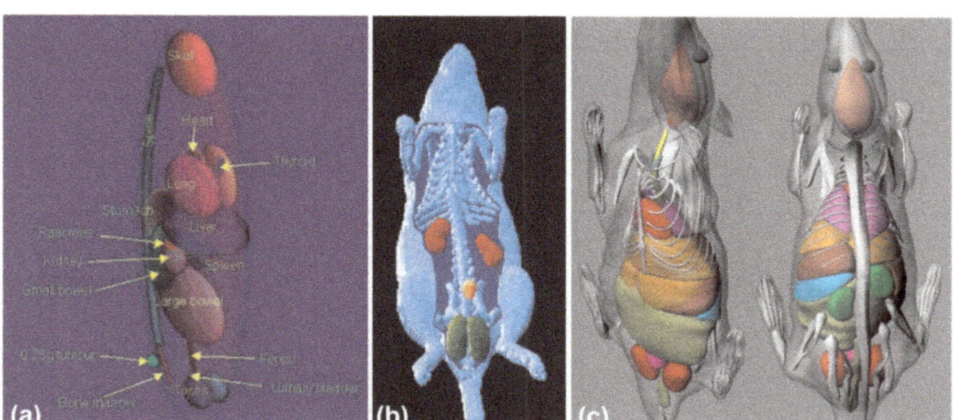

Figure 4.1. Comparison between three categories of computational models showing: (a) a stylized rat model [11], (b) a voxel-based rat model [19] and (c) the BREP rat model [23]. Reproduced with permission from [11, 19, 23].

Figure 4.1 shows three types of representative computational rat models including (a) the stylized rat model by Konijnenberg *et al* [11], (b) the voxel-based rat model by Peixoto *et al* [19] and (c) the hybrid rat model by Zhang *et al* [23]. Figure 4.2 shows the mild, moderate, severely emaciated and the obese rat models developed by Xie *et al* [30] in comparison to the normal-weight rat model. Figure 4.3 shows the ventral–dorsal views of computational rat models at different ages developed by Xie *et al* [29].

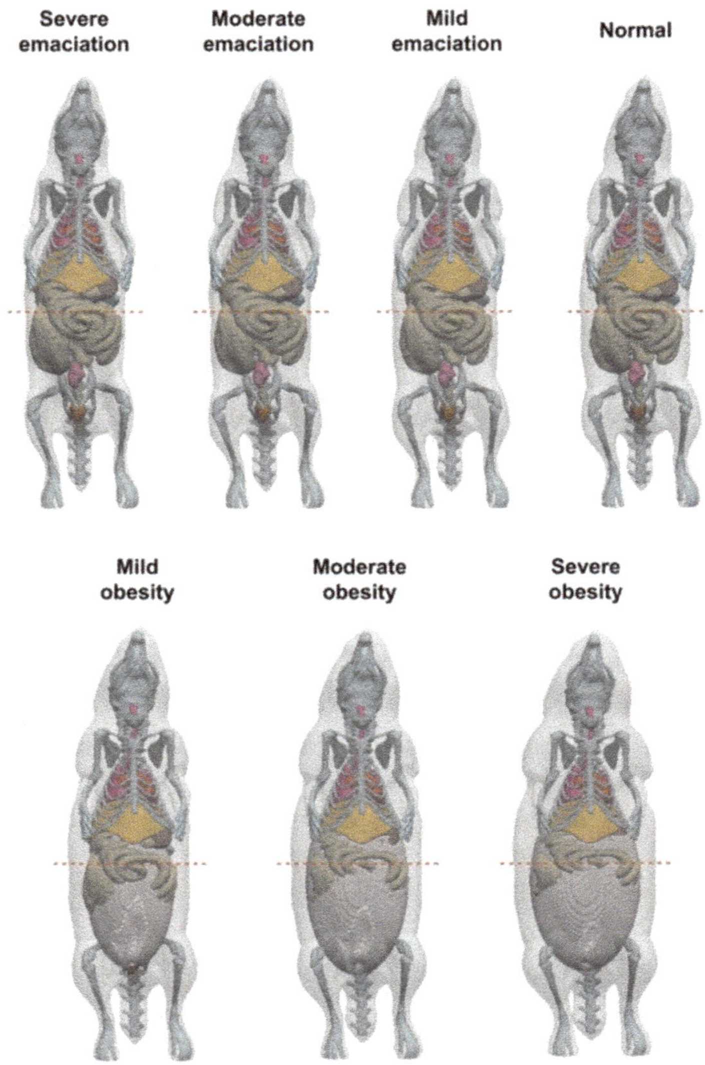

Figure 4.2. Computational rat models with different degrees of emaciation and obesity. Reproduced with permission from [30].

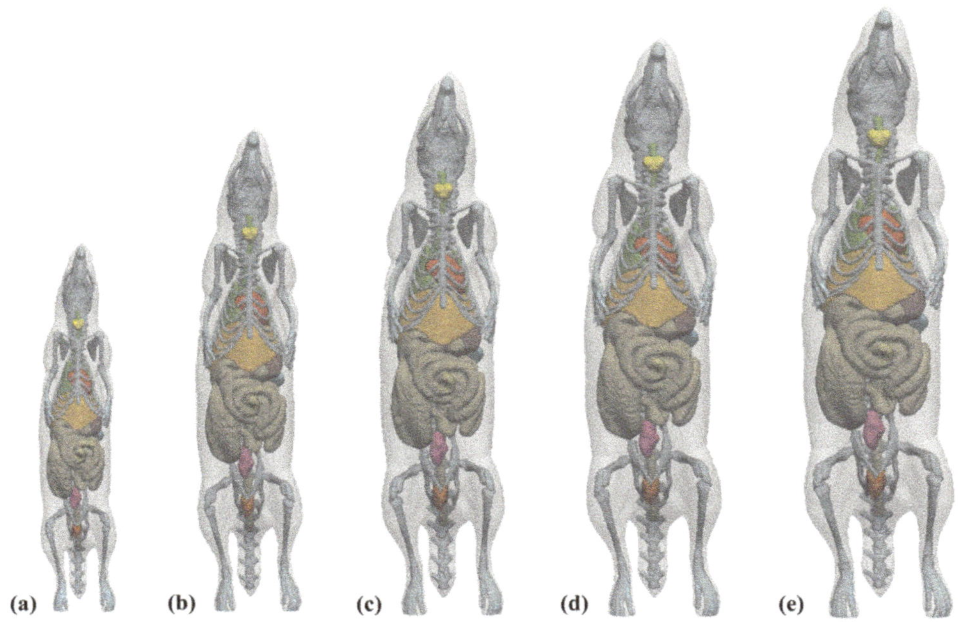

Figure 4.3. 3D visualization of computational rat models at different ages: (a) the 4 week, (b) 8 week, (c) 12 week, (d) 21 week and (e) the 45 week old models. Reproduced with permission from [29].

4.3 Development and application of HUST computational rat models

In HUST, we constructed a high-quality cryo-sectional image acquisition system for small animals and obtained a color photographic image dataset of an adult male SD rat sample with a length of 14.69 cm, height of 4.22 cm and weight of 156 g. The rat was continuously sectioned with 0.02 mm vertical intervals and photographed to obtain digital color images with 4600 × 2580 pixels and a pixel resolution of 0.2 mm × 0.2 mm. A total of 9475 slice images were generated in this procedure and were stored in TIFF format of 24 bits. The whole image set took about 6.28 GB of computer storage. The Photoshop software (Adobe Systems) was adopted for the manual segmentation and labeling process. In the constructed voxel-based rat model, 17 organs, including bladder, brain, esophagus, eyes, heart, intestine, kidney, liver, lung, skeleton, spleen, stomach, testes, thyroid, blood vessels, muscle and skin were segmented and identified. In particular, the skeleton system of the rat model was segmented into three components: the mineral bone, the red bone marrow (RBM) and the yellow bone marrow (YBM). Figure 4.4 shows volume rendering of the obtained original photographic dataset. Figure 4.5 shows one segmented slice with corresponding identified internal organs. Figure 4.6 shows the 3D whole-body distributions of mineral bone, RBM and YBM in the skeleton system.

Based on the segmented image dataset, Zhang *et al* [23] developed 3D mesh surface models for each internal organ and the body contour using the 3D-DOCTOR (Able Software Corp., Lexington, MA) software, which were next imported to

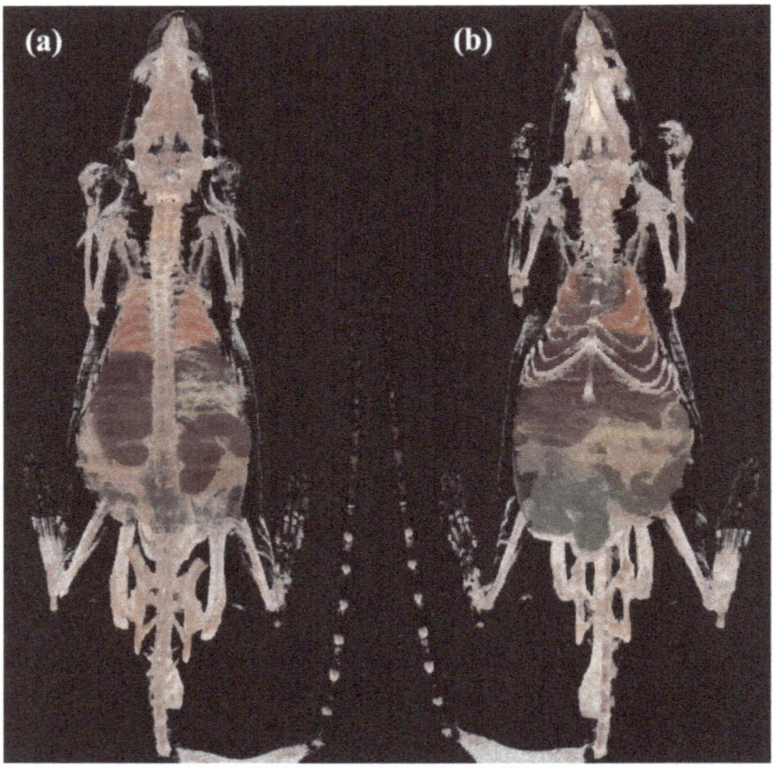

Figure 4.4. Volume rendering of the original HUST cryo-sectional photographic rat dataset: (a) dorsal–ventral view; (b) ventral–dorsal view.

Rhinoceros (McNeel North America, Seattle, WA) software in order to assemble the hybrid model as shown in figures 4.7(c) and (d). Later on, Xie *et al* [25] developed an in-house C++ program to calculate the minimum bounding box of each organ and the trunk and constructed a stylized rat phantom with 14 identified organs shown in figures 4.7(e) and (f). In the stylized rat phantom, the simple shaped organs, as well as the skeleton and trunk are conveniently expressed through mathematical equations (i.e. ellipsoid, cylinder and elliptical torus) while the relative positions and the axes of ellipsoids which describe internal organs are determined by the center and the dimensions (length, width and height) of the calculated bounding box.

The voxel-based HUST rat models were then integrated in the MCNPX Monte Carlo code to simulate the radiation transport under various exposure settings. The first application was on dosimetry. For example, the dose conversion coefficients for external photon sources, the absorbed dose fractions of internal photon and electron emitters, as well as the *S*-values for radionuclides ^{169}Er, ^{143}Pr, ^{89}Sr, ^{32}P and ^{90}Y were calculated through simulations [19, 24, 27–29].

Furthermore, the stylized HUST rat model was implemented in MCNPX-based Monte Carlo simulation code for comparing the performance of three types of computational models in radiation dosimetry. Figure 4.9 presents the self-absorbed

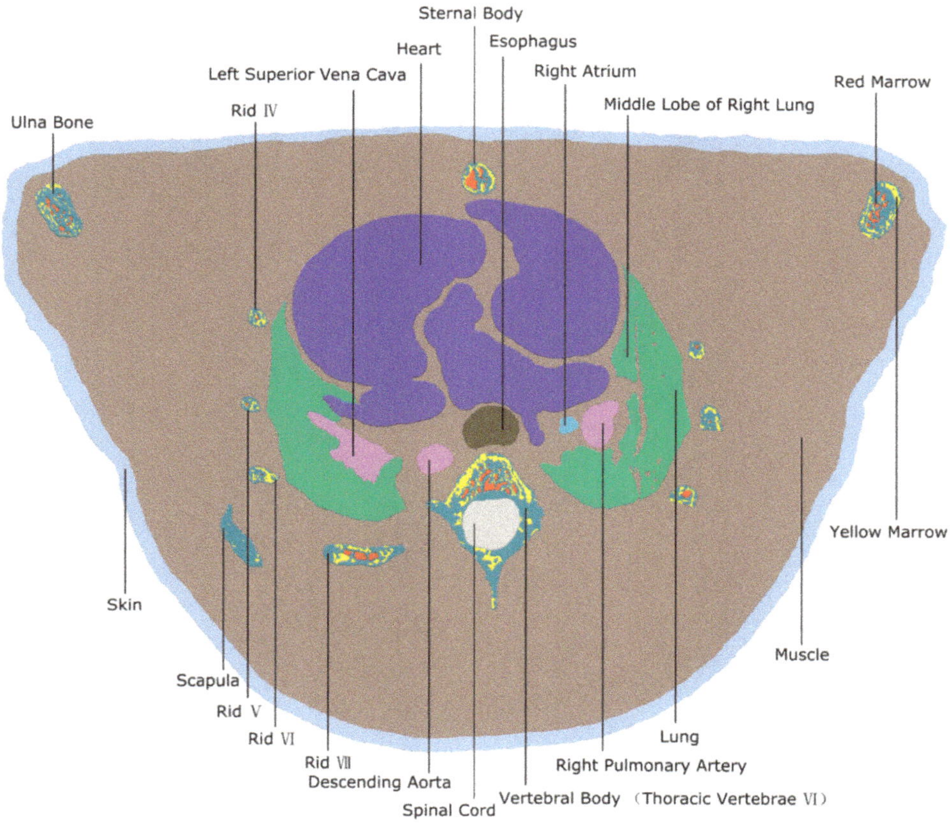

Figure 4.5. One slice of the segmented atlas of the HUST rat dataset.

photon SAF for the lungs of the three types of models. Consistent tendencies of curves are presented and most self-absorbed SAF values from these three models are close together when the photon energy is higher than 0.6 MeV. However, at the starting area of the curves, SAF values of stylized phantom show a considerable difference from the other two phantoms. The voxel-based phantom and NURBS-based phantom present consistent results in the evaluation of self-absorbed SAF while the stylized phantom results in significant divergences on estimated SAF values from low-energy photons. The hybrid model preserves most anatomic features thus being considered as the most accurate dose estimation.

Another application of the rat model is in medical imaging, which can be used to understand the physics of the imaging process, to validate novel system design, to test new algorithms and to investigate protocol settings. The hybrid rat model, for instance, was used as ground truth to evaluate the performance of the filtered back-projection algorithm on microCT, where the entire imaging chain, from acquisition to reconstruction, and all major system components, including source, projection and detector, were modeled. The reconstructed axial slices with projection data obtained by simulating a cone-beam CT system using the hybrid HUST rat model

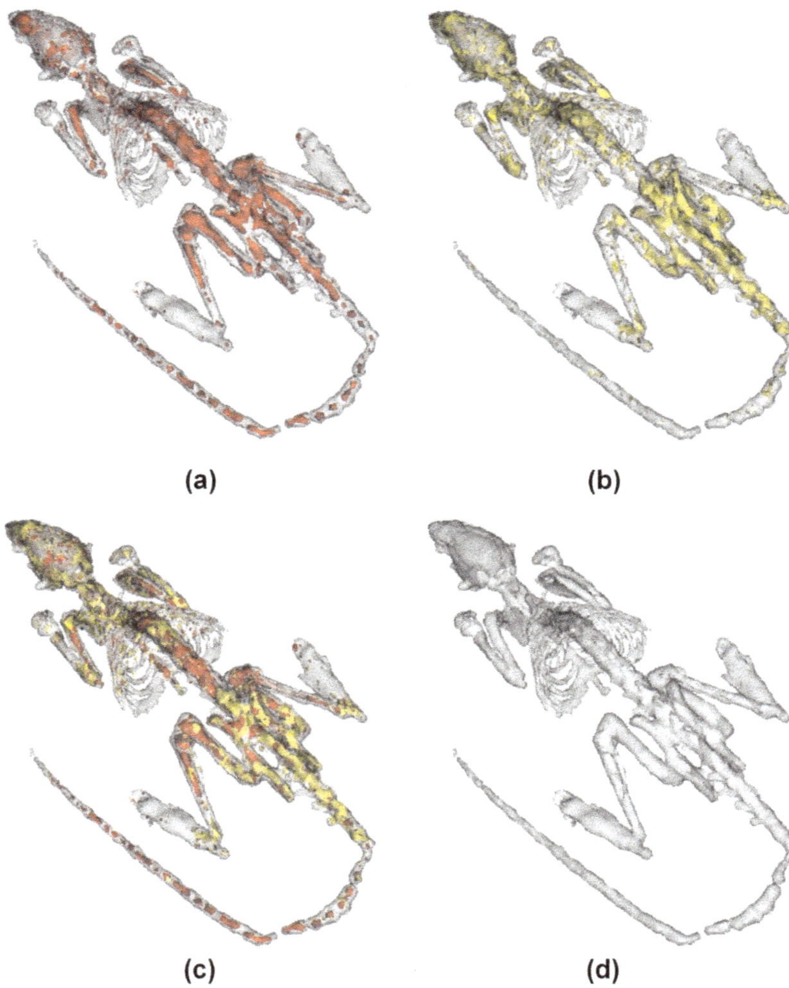

Figure 4.6. 3D visualization of whole-body distribution of mineral bone (gray), YBM (yellow) and RBM (red) in the skeleton system of the HUST rat model.

are shown in figure 4.8. It can be found that the smooth and continuous organ boundaries of the hybrid model are particularly advantageous for high-resolution imaging simulations.

Rats have been used in preclinical research for testing radiolabeled microspheres for HCC therapy where the microspheres are mainly restricted to different liver lobes [32]. However, little attention has been paid to the radiation dose in different liver lobes from radiopharmaceuticals while the whole liver was commonly served as one target region for radiation dose calculation in radiation therapy experiments with rat tumor models. In this context, we constructed a new voxel-based rat model with detailed liver lobes for a 153 g SD rat in 2010. The image dataset contains a total of 1646 successive axial color photographic images and was collected using the high-quality cryo-sectional imaging system at HUST. The original voxel dimensions

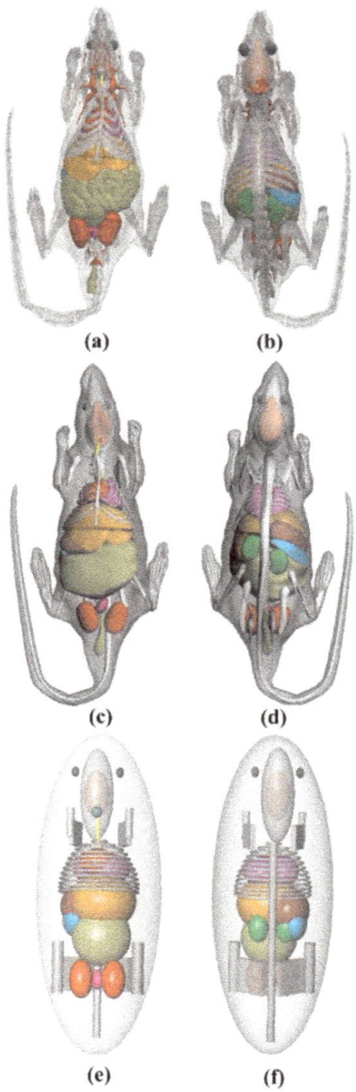

Figure 4.7. Ventral–dorsal views (1st column) and dorsal–ventral views (2nd column) of the voxel-based (top row), hybrid (middle row) and stylized rat model (bottom row) developed at HUST. Reproduced with permission from [25].

were $0.02 \times 0.02 \times 0.1$ mm^3. The left kidney, right kidney, liver, lungs, spleen, heart, stomach, brain, spinal cord, thyroid, thymus, skeleton, eyes, esophagus, pancreas and blood were identified and delineated by manual segmentation. The liver region of this model was divided into seven anatomically distinct regions: lobus sinister lateralis, lobus dexter medialis, processus caudatus, lobus sinister medialis, lobus dexter lateralis, processus papillaris I and processus papillaris II. Figure 4.10 shows one original color image in the obtained rat dataset, the segmented atlas on one slice at the level of the abdomen and the 3D visualization of the liver with identified lobes.

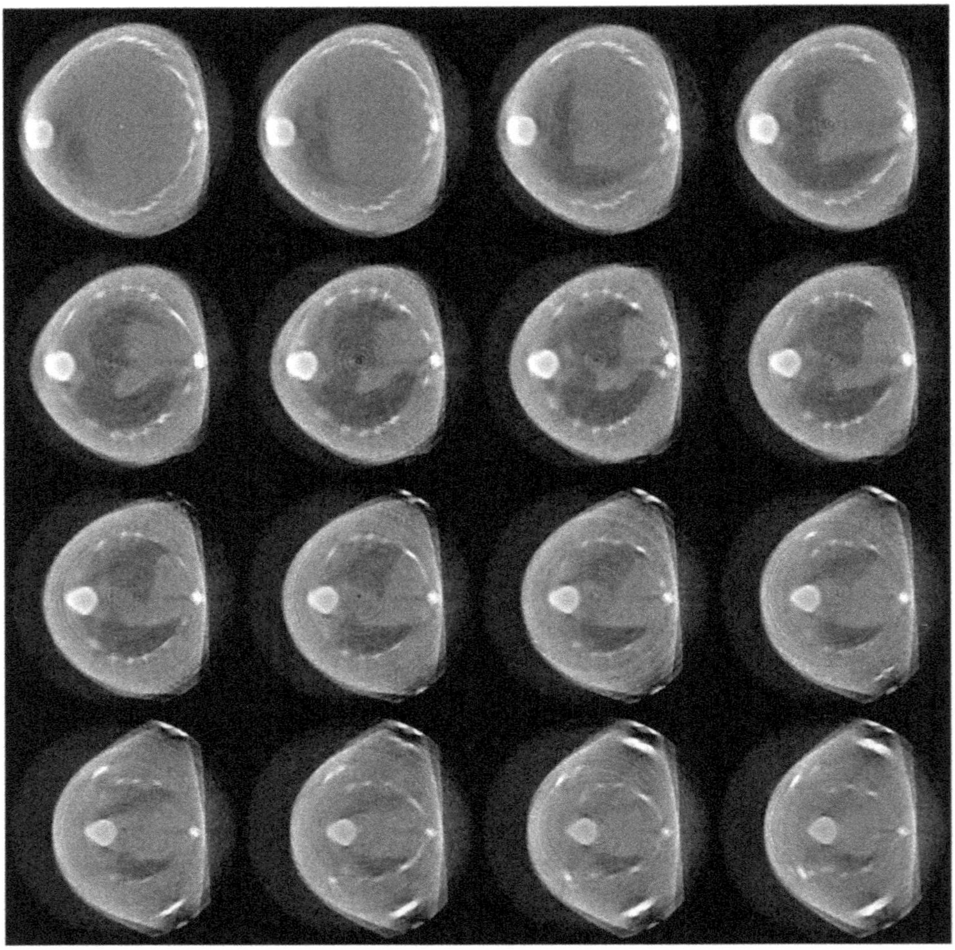

Figure 4.8. Reconstructed images of cone-beam x-ray CT simulation using the hybrid HUST rat model. Reproduced with permission from [23].

Based on this new rat model, we calculated the photon and electron AFs and compared the radiation dose for radionuclides of ^{131}I, ^{90}Y, ^{188}Re and ^{166}Ho which were reported to be used in radiation therapy of hepatocellular carcinoma (HCC). Our calculation suggests that, for absorbed dose calculation of rat liver, it is important to consider the different lobes separately for accurate dose estimation and the ^{166}Ho and ^{188}Re seem to be safer and more effective isotopes for radiation therapy of HCC comparing to ^{131}I and ^{90}Y. The developed finer liver structures can offer the opportunity to improve the accuracy of radiation dosimetry in rat liver.

4.4 Summary

Currently, more than 60 computational rat models covering three main rat strains have been reported and used in research of preclinical image processing, non-ionizing

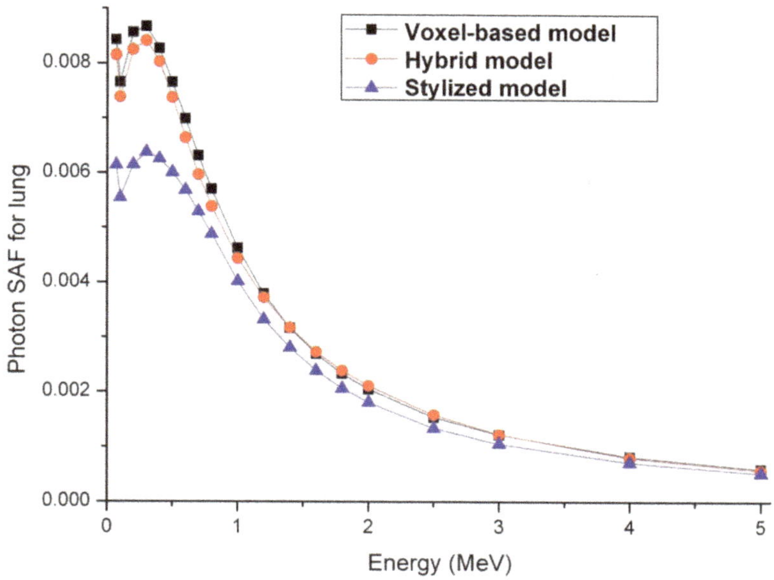

Figure 4.9. Photon self-absorbed SAF for the lungs of the voxel-based model, hybrid model and stylized model developed at HUST. Reproduced with permission from [25].

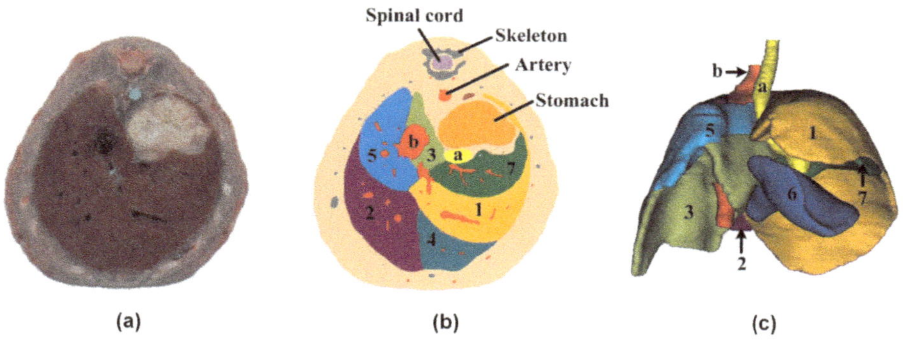

Figure 4.10. The liver model developed on the basis of the HUST cryo-sectional rat dataset. (a) Original color image of one cryo-sectional slice at the abdomen level, (b) the atlas of the same slice by manual segmentation and (c) the 3D visualization of the liver lobes, where 1 = lobus sinister lateralis (orange), 2 = lobus dexter medialis (purple), 3 = processus caudatus (green/yellow), 4 = lobus sinister medialis (teal), 5 = lobus dexter lateralis (deep skyblue), 6 = processus papillaris I (dark blue), 7 = processus papillaris II (green), a = esophagus (yellow) and b = vein (red). Reproduced with permission from [28]. © 2012 American Association of Physicists in Medicine.

and ionizing radiation dosimetry and multimodality medical physics. Among these rat models, more than 80% are voxel-based while only two hybrid models have been developed for the SD rat and the Wistar rat, respectively. It has been shown that the voxel-based models and hybrid models achieved better anatomical realism than stylized models and could be used as reliable computational models for preclinical research [24]. With the rapid development of nuclear medicine and molecular imaging technique, more and more rats are being used in the test and development of new radiotracers for targeted therapy and imaging. The anatomical characteristics (age, weight and length, etc) of the laboratory animals used would be varied. Therefore, selection of appropriate computational models would be an important issue in preclinical practices. In general, as the organ mass determines the dosimetric aspects, the computational model having the closest strain, length, age and body weight, as the experimental subject would be the most appropriate model.

References

[1] Xie T and Zaidi H 2016 Development of computational small animal models and their applications in preclinical imaging and therapy research *Med. Phys.* **43** 111–31

[2] Wang Z, Yang W L, Jacob A, Aziz M and Wang P 2015 Human ghrelin mitigates intestinal injury and mortality after whole body irradiation in rats *PloS one* **10** e0118213

[3] E Robbins M, Zhao W, A Garcia-Espinosa M and I Diz D 2010 Renin-angiotensin system blockers and modulation of radiation-induced brain injury *Curr. Drug Targets* **11** 1413–22

[4] Jourdan M, Lopez A, Olasz E, Duncan N, Demara M and Kittipongdaja W *et al* 2011 Laminin 332 deposition is diminished in irradiated skin in an animal model of combined radiation and wound skin injury *Radiat. Res.* **176** 636–48

[5] Burkhardt M, Spinelli Y and Kuster N 1997 Exposure setup to test effects of wireless communications systems on the CNS *Health Phys.* **73** 770–8

[6] Lapin G and Allen C 1997 Requirements for accurate anatomical imaging of the rat for electromagnetic modeling *Proc. of the 19th Annual Int. Conf. of the IEEE Engineering in Medicine and Biology Society* IEEE

[7] Mason P, Ziriax J, Hurt W and D'Andrea J 1999 3-dimensional models for EMF dosimetry *Electricity and Magnetism in Biology and Medicine* (Berlin: Springer) pp 291–4

[8] Chou C, Chan K, McDougall J and Guy A 1999 Development of a rat head exposure system for simulating human exposure to RF fields from handheld wireless telephones *Bioelectromagnetics* **20** 75–92

[9] Watanabe S, Mukoyama A, Wake K, Yamanaka Y, Uno T and Taki M 2000 Microwave exposure setup for a long-term *in vivo* study *Proc. of the Int. Symp. on Antennas and Propagation*

[10] Funk T, Sun M and Hasegawa B H 2004 Radiation dose estimate in small animal SPECT and PET *Med. Phys.* **31** 2680–6

[11] Konijnenberg M W, Bijster M, Krenning E P and de Jong M 2004 A stylized computational model of the rat for organ dosimetry in support of preclinical evaluations of peptide receptor radionuclide therapy with 90Y, 111In, or 177Lu *J. Nucl. Med.* **45** 1260–9

[12] Leveque P, Dale C, Veyret B and Wiart J 2004 Dosimetric analysis of a 900-MHz rat head exposure system *IEEE Trans Microw Theor. Tech.* **52** 2076–83

[13] Schönborn F, Poković K and Kuster N 2004 Dosimetric analysis of the carousel setup for the exposure of rats at 1.62 GHz *Bioelectromagnetics* **25** 16–26

[14] Segars W P, Tsui B M, Frey E C, Johnson G A and Berr S S 2004 Development of a 4-D digital mouse phantom for molecular imaging research *Mol. Imaging Biol.* **6** 149–59

[15] Stabin M G, Peterson T E, Holburn G E and Emmons M A 2006 Voxel-based mouse and rat models for internal dose calculations *J. Nucl. Med.* **47** 655–9

[16] Kainz W, Nikoloski N, Oesch W, Berdiñas-Torres V, Fröhlich J and Neubauer G *et al* 2006 Development of novel whole-body exposure setups for rats providing high efficiency, National Toxicology Program (NTP) compatibility and well-characterized exposure *Phys. Med. Biol.* **51** 5211

[17] Lopresto V, Pinto R, De Vita A, Mancini S, Galloni P and Marino C *et al* 2007 Exposure setup to study potential adverse effects at GSM 1800 and UMTS frequencies on the auditory systems of rats *Radiat. Prot. Dosim.* **123** 473–82
[18] ICRP. Publication 108 2008 Environmental protection: The concept and use of reference animals and plants *Ann. ICRP* **38**
[19] Peixoto P, Vieira J, Yoriyaz H and Lima F 2008 Photon and electron absorbed fractions calculated from a new tomographic rat model *Phys. Med. Biol.* **53** 5343
[20] Wu L, Zhang G, Luo Q and Liu Q 2008 An image-based rat model for Monte Carlo organ dose calculations *Med. Phys.* **35** 3759–64
[21] Wang J, Fujiwara O, Wake K and Watanabe S 2008 Dosimetry evaluation for pregnant and fetus rats in a near-field exposure system of 1.95-GHz cellular phones *IEEE Microw. Wirel. Compon. Lett.* **18** 260–2
[22] Pain F, Dhenain M, Gurden H, Routier A, Lefebvre F and Mastrippolito R *et al* 2008 A method based on Monte Carlo simulations and voxelized anatomical atlases to evaluate and correct uncertainties on radiotracer accumulation quantitation in beta microprobe studies in the rat brain *Phys. Med. Biol.* **53** 5385
[23] Zhang G, Xie T, Bosmans H and Liu Q 2009 Development of a rat computational phantom using boundary representation method for Monte Carlo simulation in radiological imaging *Proc. IEEE* **97** 2006–14
[24] Keenan M A, Stabin M G, Segars W P and Fernald M J 2010 RADAR realistic animal model series for dose assessment *J. Nucl. Med.* **51** 471–6
[25] Xie T, Zhang G, Li Y and Liu Q 2010 Comparison of absorbed fractions of electrons and photons using three kinds of computational phantoms of rat *Appl. Phys. Lett.* **97** 033702
[26] Arima T, Watanabe H, Wake K, Masuda H, Watanabe S and Taki M *et al* 2011 Local exposure system for rats head using a figure-8 loop antenna in 1500-mhz band *IEEE Trans. Biomed. Eng.* **58** 2740–7
[27] Larsson E, Ljungberg M, Mårtensson L, Nilsson R, Tennvall J and Strand S-E *et al* 2012 Use of Monte Carlo simulations with a realistic rat phantom for examining the correlation between hematopoietic system response and red marrow absorbed dose in Brown Norway rats undergoing radionuclide therapy with 177Lu-and 90Y-BR96 mAbs *Med. Phys.* **39** 4434–43
[28] Xie T, Liu Q and Zaidi H 2012 Evaluation of S-values and dose distributions for 90Y, 131I, 166Ho, and 188Re in seven lobes of the rat liver *Med. Phys.* **39** 1462–72
[29] Xie T and Zaidi H 2013 Age-dependent small-animal internal radiation dosimetry *Mol. Imaging* **12**
[30] Xie T and Zaidi H 2013 Effect of emaciation and obesity on small-animal internal radiation dosimetry for positron-emitting radionuclides *Eur. J. Nucl. Med. Mol. Imaging* **40** 1748–59
[31] Locatelli M, Miloudi H, Autret G, Balvay D, Desbrée A and Blanchardon E *et al* 2017 RODES software for dose assessment of rats and mice contaminated with radionuclides *J. Radiol. Prot.* **37** 214
[32] Nijsen F, Rook D, Brandt C, Meijer R, Dullens H and Zonnenberg B *et al* 2001 Targeting of liver tumour in rats by selective delivery of holmium-166 loaded microspheres: A biodistribution study *Eur. J. Nucl. Med. Mol. Imaging* **28** 743–9

IOP Publishing

Computational Anatomical Animal Models
Methodological developments and research applications
Habib Zaidi

Chapter 5

Overview of computational frog models

Sakae Kinase, Akram Mohammadi and José-María Gómez-Ros

5.1 Introduction

The International Commission on Radiological Protection (ICRP) has made general statements about the protection of the environment in 2007, recommendations published as ICRP Publication 103 [1]. The importance of radiation dosimetry on non-human biota has attracted growing interest from the viewpoint of environmental protection. To form the basis of a more structured approach to understanding the relationships between radiation exposures and dose, between dose and radiological effects, and between effect and possible consequences, ICRP proposes the use of reference animals and plants (RAPs): deer, rats, ducks, frogs, trout, flatfish, bees, crabs, earthworms, pine trees, grass and brown seaweed [2–4]. ICRP has evaluated dosimetric quantities on RAPs such as the reference frog—absorbed fractions (AFs), which are defined as the fraction of energy emitted by a radiation source that is absorbed within the target tissue, organ or organism. However, there is a great lack of reliable data on the AFs for the organ dose evaluations since ICRP assessed the AFs for the whole organisms using simplified dosimetric models such as ellipsoids and cylinders—stylized models. The stylized model for the reference frog remains simplified as the shape is based only on the whole body. Kinase focused his attention on frogs that take many exposed situations such as in soil, on soil and in water, and developed a realistic frog model for radiation protection of the environment [5]. The realistic frog model is defined using digital volume arrays—voxel-based models. The voxel-based frog model is represented by matrices of segmented voxels from cryosection data. Dosimetric quantities for the organs of the voxel-based frog model have been evaluated using Monte Carlo simulations [5–10]. Table 5.1 summarizes the computational frog models, where anatomical features describe the characteristics of used frog specimens, the model number is the number of unique frog models reported in the referenced publication, and imaging refers to the imaging technique from which the model is derived [11].

doi:10.1088/2053-2563/aae1b4ch5

Table 5.1. Summary of developed computational frog models [11].

Year	Developer	Affiliation	Anatomical features	Model number	Imaging modality	Model type	Applications
2008	ICRP [2]	ICRP	31.4 g frog	1	Anatomic data	Stylized model	Dose conversion factors for frogs under internal/ external exposure to 75 radionuclides
2008	Kinase [5]	JAEA, Japan	33.7 g frog	1	Cryosection data	Voxel model	Radiation dose of monoenergetic electrons and photons

To the authors' best knowledge, there are limited investigations on the computational frog models and the organ dose evaluations for frogs in environmental protection. In this chapter, computational frog models and their applications are reviewed to share some perspectives of frog model development in the near future.

5.2 History and construction of computational frog models

In 2008, ICRP recommended an environmental protection framework based on 12 RAPs such as the reference frog to which dosimetric quantities can be evaluated using simplified models. The assumed shapes and dimensions of the reference frog are given in ICRP Publication 108 [2]. The individual egg of the reference frog has a mass of 5.24×10^{-4} kg, and is represented by a sphere with a diameter of 1 cm. A 'clump' of spawn of the reference frog has a mass of 0.314 kg, and is represented by an ellipsoid with dimensions of $20 \times 6 \times 5$ cm. A tadpole of the reference frog has a mass of 4.42×10^{-4} kg, and is represented by an ellipsoid with dimensions of $1.5 \times 0.75 \times 0.75$ cm. An adult reference frog has a mass of 3.14×10^{-2} kg, and is represented by an ellipsoid with dimensions of $8 \times 3 \times 2.5$ cm. The dosimetric models for the reference frog are based on whole-body geometry alone and do not include internal organs.

In 2008, the voxel-based frog model was developed based on the segmented images from a frog [5]. The original segmented images which are constructed from cryosection data, have been available on the website of the Lawrence Berkeley National Laboratory [12]. The voxel-based frog model has a mass of 3.37×10^{-2} kg. The dimensions are 7.1 (length) × 3.3 (width) × 2.4 (height) cm. The size of the voxel-based frog model corresponds to that of the ICRP adult reference frog. The organ/tissue volumes are summarized in table 5.2. Figure 5.1 shows cross sections from the voxel-based frog model. The data set consists of 138 slices of 470 × 500 pixels, with a voxel size 175 μm. Every voxel was assigned a specific organ identification (ID) number. Each organ/tissue was represented by voxels identified as belonging to it from the cryosection data.

Figure 5.2 shows the ICRP adult reference frog and the voxel-based frog model. The voxel-based frog model provides better anatomical realism than the ICRP adult reference frog. To obtain more sophisticated hybrid models which are capable of deforming the organs and body, advanced boundary representation techniques such as the B-splines, NURBS, and polygon meshes would be used.

5.3 Monte Carlo simulations with computational frog models

Dosimetric quantities such as AFs, *S* values, dose conversion coefficients (DCCs) or dose conversion factors (DCFs) for computational frog models have been evaluated using Monte Carlo simulations. The popular Monte Carlo codes such as EGS [13] and MCNP [14] were used to simulate radiation transport inside the frog body, thus providing patterns of radiation interactions and estimated distributions of energy deposition.

Table 5.2. Organ/tissue volumes and masses of the voxel-based frog model. Reprinted with permission from [5].

Organ	Volume (cm^3)	Organ mass (kg)
Brain	1.03×10^{-1}	1.07×10^{-4}
Blood vessel	1.88×10^{-1}	1.95×10^{-4}
Duodenum	2.05×10^{-1}	2.13×10^{-4}
Eye	2.85×10^{-1}	2.97×10^{-4}
Heart	1.99×10^{-1}	2.07×10^{-4}
Ileum	1.49×10^{-1}	1.55×10^{-4}
Kidneys	2.10×10^{-1}	2.19×10^{-4}
Intestine	2.75×10^{-1}	2.86×10^{-4}
Liver	1.10	1.14×10^{-3}
Lung	2.54×10^{-1}	7.53×10^{-5}
Soft tissue	2.59×10^{1}	2.70×10^{-2}
Nerve	1.31×10^{-1}	1.36×10^{-4}
Skeleton	2.36	3.31×10^{-3}
Spleen	2.11×10^{-2}	2.19×10^{-5}
Stomach	1.32	1.37×10^{-3}

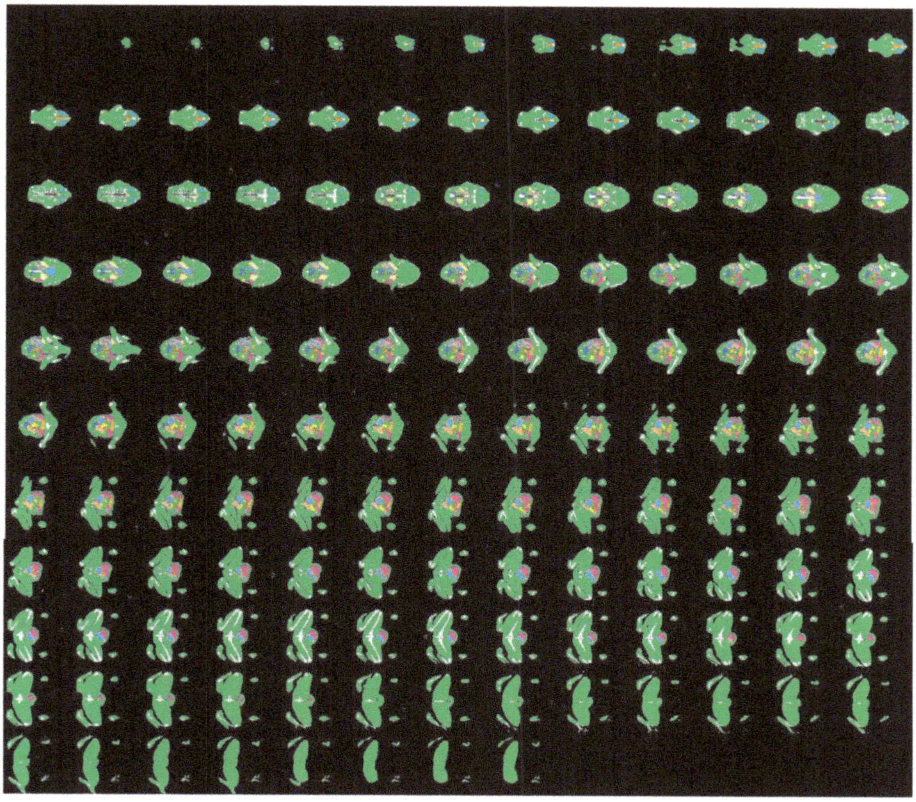

Figure 5.1. Cross sections of the voxel-based frog model. Reprinted with permission from [6].

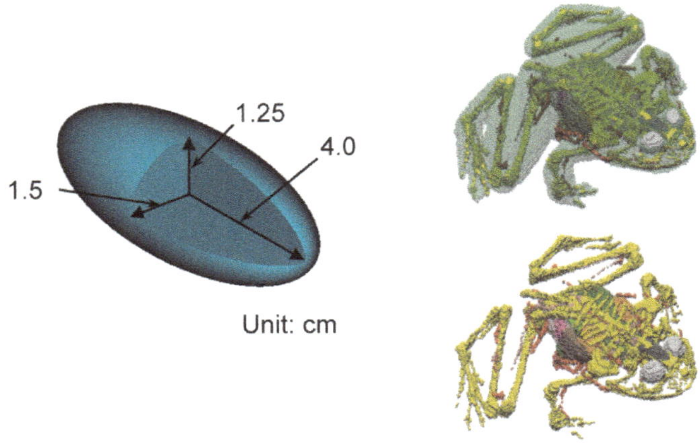

(a) (b)

Figure 5.2. (a) The ICRP adult reference frog and (b) the voxel-based frog models [10].

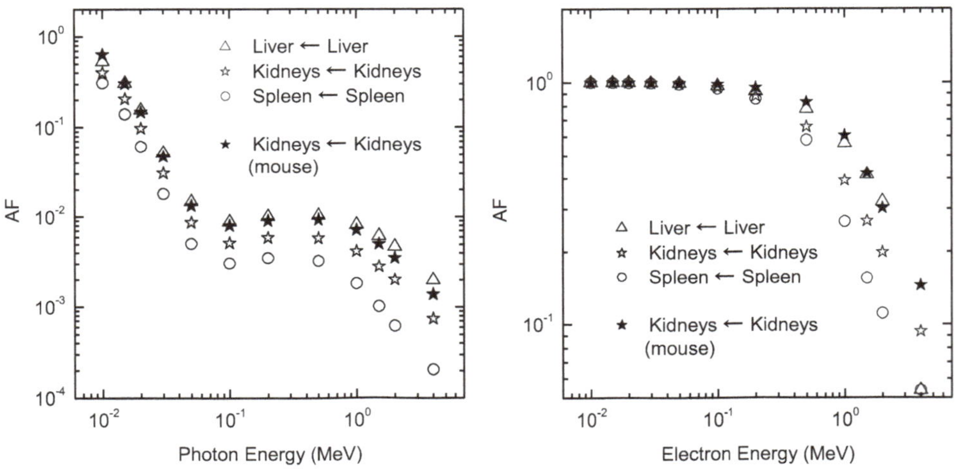

Figure 5.3. Self-AFs in the spleen, kidneys and liver of the voxel-based frog model in the photon/electron energy 0.01–4 MeV. The self-AFs in the kidneys of the mouse model are plotted. Reprinted with permission from [5].

5.3.1 Absorbed fractions and *S* values for the voxel-based model

Self-absorbed fractions (AFs) for photons and electrons in the spleen, kidneys and liver of the voxel-based frog model were evaluated using the EGS4 [13], in conjunction with an EGS4 user code, UCSAF [15]. The geometry of the EGS4-UCSAF code treats voxel models. The self-AFs were evaluated for uniformly distributed monoenergetic sources of both photons and electrons at discrete initial energies: 0.01, 0.015, 0.02, 0.03, 0.05, 0.1, 0.2, 0.5, 1.0, 1.5, 2.0 and 4.0 MeV. Photon and electron histories were run at numbers sufficient to reduce statistical uncertainties to below 5%. Figure 5.3 shows the self-AFs for photons and electrons in the spleen, kidneys and liver of the voxel-based frog model in the energy range from 0.01 to 4 MeV. The self-AFs for photons decrease with an increase in photon energy on the whole. The self-AFs for electrons are almost unity in the electron energy range from 0.01 to 0.1 MeV, followed by a sharp fall. In the figure, the self-AFs for photons and electrons in the kidneys (5.15×10^{-4} kg) of the voxel-based mouse model [16] were also plotted for comparison. The self-AFs for the voxel-based frog model are consistent with those for the mouse model. This is due to the fact that the self-AFs depend on the organ mass.

Self-*S* values (μGy MBq^{-1} s^{-1}) for the spleen, kidneys and liver in the voxel-based frog model were calculated using the results of the self-AFs for both photons and electrons. The nuclides considered were ^{18}F and ^{90}Y of potential interest in the radioactive medical wastes. Figure 5.4 shows the self-*S* values for the spleen, kidneys and liver of the voxel-based frog model, for ^{18}F and ^{90}Y. It can be seen that, the self-*S* values for the voxel-based frog model are dependent on the organ mass.

5.3.2 Dose coefficients (DCs) for the voxel-based model

Dose conversion coefficients (DCCs) (μGy h^{-1} Bq^{-1} kg) for the assessment of internal absorbed dose rate in the voxel-based frog model were evaluated using the MCNPX [9, 10, 17]. The source of the photons and electrons was assumed to be monoenergetic in the energy range 0.01–5 MeV, uniformly distributed in the source organ/tissue of the voxel-based frog model. The source organs for photons and electrons were the brain, liver, lung, skeleton, stomach and whole body. The target was the whole body. A cut-off in the number of histories was applied to obtain relative errors less than 3%. Figure 5.5 shows the energy dependence of the DCCs for photons and electrons in the voxel-based frog model. The DCCs for photons in

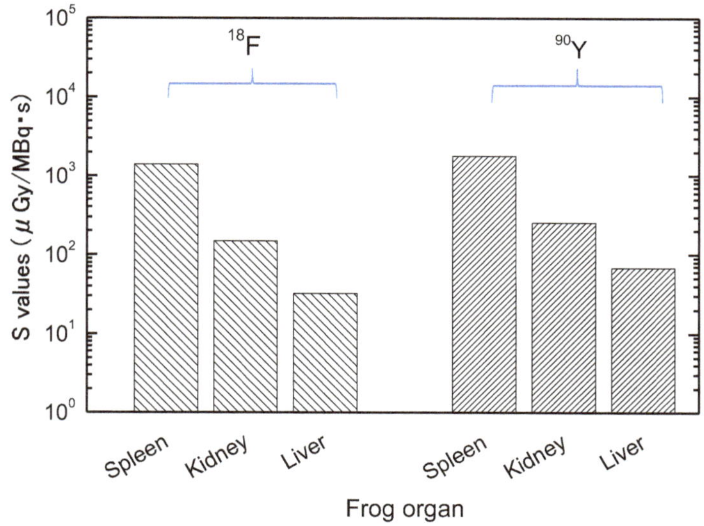

Figure 5.4. Self-*S* values for ^{18}F and ^{90}Y in the spleen, kidneys and liver of the voxel-based frog model. Reprinted with permission from [5].

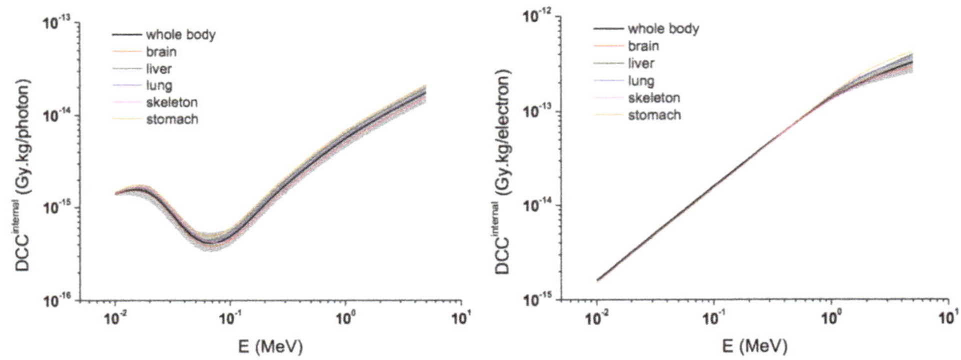

Figure 5.5. Dose conversion coefficients (DCCs) for the voxel-based frog model in the photon/electron energy 0.01–5 MeV. The target is the whole body [10].

the voxel-based frog model are not very dependent on the energy except for energies below about 0.05 MeV, when the photon mean free path in organ/tissue is much less than the size of the frog model. The DCCs for electrons increase with an increase in electron energy on the whole. Gómez-Ros *et al* reported that uncertainties of the whole-body DCC due to non-homogeneous radionuclide distribution are less than 30% for photons and electrons [9]. In ICRP Publication 108, the term 'dose coefficients (DCs)' replaces the term 'dose conversion coefficients (DCCs)' and 'dose conversion factors (DCFs)'.

5.3.3 Comparisons between stylized and voxel-based models

Photon and electron AFs in the whole body of the voxel-based frog model were evaluated using the EGS4-UCSAF code and compared with those in the ICRP adult reference frog [7, 8]. The voxel size of the original voxel-based frog model was changed from 0.035 to 0.034 cm to have the same volume as the ICRP adult reference frog. The dimensions of the ICRP adult reference frog were slightly changed to convert it to a voxelized ellipsoid with the voxel size of 0.034 cm. The ICRP adult reference frog was re-constructed as an ellipsoid with dimensions of $8.024 \times 2.992 \times 2.516$ cm. The volumes of the voxel-based frog model and the ICRP adult reference frog were 31.52 cm^3 and 31.62 cm^3, respectively. Photon and electron AFs in the whole body of the voxel-based frog model and the ICRP adult reference frog are shown in figure 5.6. Electron AFs in the voxel-based frog model and the ICRP adult reference frog are almost the same for electrons with energy less than 0.5 MeV since electron range is not comparable with the dimensions of the models. Differences between photon AFs in the two models are 24% at maximum, while for electron AFs they are less than 19%. The whole-body shape is the important parameter which can influence the AFs since the two models have the same masses. Hence, AFs depend on the whole-body shape and considering only the mass as a parameter which can affect AFs causes large uncertainty. Mohammadi *et al* also reported that whole-body *S* values in simplified models without internal organs are

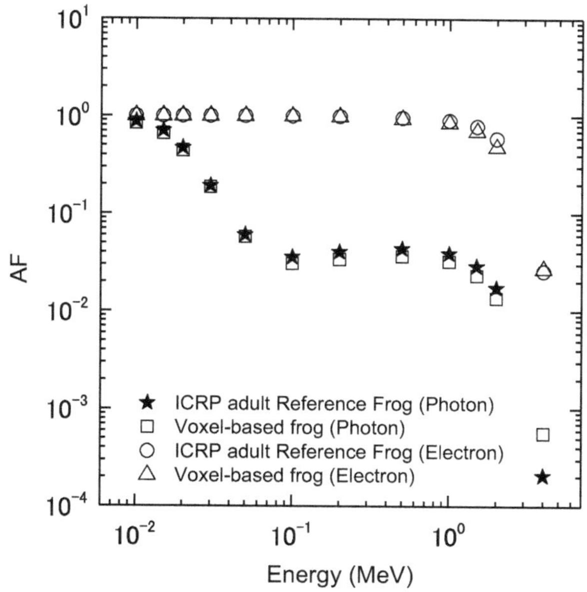

Figure 5.6. Self-AFs in the whole body of the voxel-based frog model and the ICRP adult reference frog in the photon/electron energy 0.01–4 MeV.

not sufficient for accurate internal dosimetry since they do not reflect *S* values of all individual organs as the source was not distributed uniformly in the whole body [8].

5.4 Summary

The computational frog models have been developed for the evaluation of the exposure of non-human biota to ionizing radiation. The dosimetric quantities for photons and electrons in organ/tissue of the computational frog models have been evaluated using Monte Carlo simulations. The voxel-based frog model is useful since it provides better anatomical realism than the ICRP adult reference frog. It is confirmed that the self-AFs and self-*S* values for the voxel-based frog model are largely dependent on the mass of the source organ. The whole-body AFs in the voxel-based and the reference frog models were different by 24%, which demonstrated the effect of the whole-body shapes on the whole-body AFs. The whole-body shape impacted on AFs significantly, which may support the idea for the replacement of the ICRP adult reference animal models by voxel-based animal models to improve the accuracy of the whole-body AF. It could be argued that the use of the average whole-body activity concentration and the homogeneous DCCs to calculate the whole-body doses seem to be accurate enough for the purposes of environmental protection.

The authors believe that 3D printing frog phantoms with adequate tissue substitutes should be developed for the validation of the dosimetric quantities by the Monte Carlo simulations.

References

[1] ICRP. Publication 103: 2007 The Recommendation of the International Commission on Radiological Protection *Ann. ICRP* **37**

[2] ICRP. Publication 108: 2008 Environmental Protection: the Concept and Use of Reference Animals and Plants *Ann. ICRP* 38

[3] ICRP. Publication 114: 2009 Environmental Protection: Transfer Parameters for Reference Animals and Plants *Ann. ICRP* **39**

[4] ICRP. Publication 136: 2017 Dose Coefficients for Non-Human Biota Environmentally Exposed to Radiation *Ann. ICRP* **46**

[5] Kinase S 2008 Voxel-Based Frog Phantom for Internal Dose Evaluation *J. Nucl. Sci. Technol.* **45** 1049–52

[6] Kinase S 2009 Monte Carlo Simulations of photon absorbed fractions in a frog voxel phantom *Proc. IEEE* **97** 2086–97

[7] Mohammadi A, Kinase S and Saito K 2011 Comparison of photon and electron absorbed fractions in voxel-based and simplified phantoms for small animals *Prog. Nucl. Sci. Tech.* **2** 365–8

[8] Mohammadi A, Kinase S and Saito K 2012 Evaluation of absorbed doses in voxel-based and simplified models for small animals *Radiat. Prot. Dosim.* **150** 283–91

[9] Gómez-Ros J M, Prohl G, Ulanovsky A and Lis M 2008 Uncertainties of internal dose assessment for animals and plants due to non-homogeneously distributed radionuclides *J. Environ. Radioact.* **99** 1449–55

[10] Gómez-Ros J M 2008 Phantoms for non-human biota for radioecology *EURADOS 11th Winter School (Lisbon, Portugal, 8 February 2018)*

[11] Xie T and Zaidi H 2016 Development of computational small animal models and their applications in preclinical imaging and therapy research *Med. Phys.* **43** 111–31

[12] Johnson W and Robertson D *Whole frog project* http://froggy.lbl.gov/

[13] Nelson W R, Hirayama H and Rogers D W O 1985 The EGS4 Code System. SLAC-265

[14] Pelowitz D B 2005 MCNPX User's Manual Version 2.5.0. Report LA-CP-05-0369

[15] Kinase S, Zankl M, Kuwabara J, Sato K, Noguchi H, Funabiki J and Saito K 2003 Evaluation of specific absorbed fractions in voxel phantoms using monte carlo simulation *Radiat. Prot. Dosim.* **105** 557–63

[16] Kinase S, Takahashi M and Saito K 2008 Evaluation of self-absorbed doses for the kidneys of a voxel mouse *J. Nucl. Sci. Technol.* Suppl. 5 268–70

[17] Stark K *et al* 2017 Dose assessment in environmental radiological protection: State of the art and perspectives *J. Environ. Radioact.* **175-176** 105–14

IOP Publishing

Computational Anatomical Animal Models
Methodological developments and research applications
Habib Zaidi

Chapter 6

Overview of computational canine models

Choonsik Lee

6.1 Introduction

Animal research is crucial to the understanding of the variables involved in the behavior or biological system of humans. The results obtained from animal research are extrapolated to humans with potential uncertainty involved. Although different kinds of non-human subjects have been used in research ranging from small animals, such as rats and mice, to bigger animals such as primates, canines have several advantages against other alternatives: they are readily available; they have optimum life span, 10–15 years; and their size is closer to that of humans. Furthermore, the completion of the canine genome project confirmed that there are similarities in cancers between canines and humans [1–3].

Computational anatomy models have been widely used in radiation dosimetry studies to simulate the behavior of radiations within the computational representation of anatomies. Computational human models have been used in a variety of radiation dosimetry studies since the 1960s [4]. In radiopharmaceutical research where animal experiments are conducted, it is important to have corresponding computational models of animals available. Compared to the history and variety of computational human models, canine models have not been vastly investigated. There are only six canine models reported to date. This chapter summarizes the general steps taken in the development of computational canine models and the current status of canine models that are available presently.

6.2 General steps for developing canine models

Since developing canine models take similar steps to the development of computational human models, it is useful to provide a brief history of computational human models [4, 5]. The first generation of computational models, called stylized or mathematical models, were defined using simple mathematical equations to represent different anatomical structures. Though these stylized models offered great flexibility through their ability to easily rotate, reposition, and scale, they suffered from a lack

doi:10.1088/2053-2563/aae1b4ch6

of anatomical realism. Voxel (volumetric pixel) or tomographic models, the second generation of computational human models, are defined as a three-dimensional matrix of voxels developed from segmented radiologic images. Though much more realistic than the previous stylized models, voxel models are fundamentally fixed to the original source images of a specific patient; the shape, size, and location of organs and tissues within voxel models cannot be easily modified or deformed. These shortcomings resulted in the creation of the third and latest generation of human models, known as the hybrid or boundary representation (BREP) models, which are represented by a polygon mesh or non-uniform rational B-spline (NURBS). Hybrid models capture the best advantages of previous stylized and voxel models in terms of anatomical realism and flexibility.

Because the first canine model was introduced in 2008 [6], the modeling technique skipped the first- and second-generation computational models and directly took advantage of the latest modeling techniques. Canine models are usually developed in the following three steps: acquisition of tomographic images of canine subjects; segmentation of organs and tissues of interests from the images; and developing three-dimensional (3D) models from the organ and tissue contours segmented from images.

6.2.1 Acquisition of tomographic images

Other than the first-generation computational models, the second- and third-generation computational models are based on tomographic medical images to achieve anatomical realism. The development of the third-generation canine models starts from the acquisition of tomographic images of canine subjects. Imaging canines and further development of canine models from the images need to be approved by an appropriate ethics committee at a given institution and to be conducted strictly in accordance with proper protocols. Anesthesia is usually induced intravenously to obtain high quality images by avoiding unexpected movement of subjects. Depending on the purpose of model development, partial body or whole-body images are acquired. Computed tomography (CT) scans are usually adopted, as scanning is fast and both soft tissues and skeleton can be easily identified in the segmentation process.

6.2.2 Segmentation of organs and tissues

Once tomographic images of subjects are available, organs and tissues of interest need to be segmented from the images. Since organs and tissues must be manually segmented except for high-contrasted tissues such as the lungs and skeletons, this step could take the most of time in model development. Computer programs such as 3D-Doctor™ (Able Software Corp, Lexington, MA) are employed at this step that can visualize cross sectional images slice by slice and enable a researcher to draw the boundaries of organs and tissues on the images. Different tags or indices are assigned to different organs and tissues for the purpose of 3D modeling and radiation dosimetry. If the second-generation voxel models are aimed, the development process is completed at this step. The boundaries of organs and tissues indexed with different tag numbers are converted into binary file format for radiation dosimetry simulations. However, if the third-generation models are intended, one more step needs to be conducted as described in the following section.

6.2.3 Development of 3D whole body models

In the process of manual segmentation, several issues can occur as follows, which potentially can affect resulting radiation dose estimations. Because organs and tissues are segmented on axial (or transversal) plane of CT images, it is very difficult to make the surface of the combined 3D object smooth. Most of the second-generation voxel models suffer from this issue. Sometimes it is required to adjust the volume and/or location of organs to match them with standard anatomical data such as reference organ mass. Since the tomographic images are obtained from patients potentially having abnormal organ structures, it is often necessary to adjust the shape and location of organs. These issues can be overcome using the third-generation modeling process based on the polygon mesh or NURBS surface.

From the segmented organ boundaries in the previous step, polygon mesh models can be transferred to other computer programs that can readily handle 3D objects in polygon mesh or NURBS format such as Rhinoceros™ (McNeel North America, Seattle, WA). After finishing the adjustments mentioned above, the final organ models are combined and converted into a whole-body model in voxel format by using a process called voxelization [7]. This process was required for most Monte Carlo transport codes to simulate the behavior of radiations in media. However, more recently, it became possible to directly transport radiations within the polygon mesh models in most major Monte Carlo codes including MCNP6 [8], GEANT4 [9], and PHITS [10].

6.3 Current status of canine models

6.3.1 The University of Florida canine models

The researchers at the University of Florida (UF) first introduced a canine model [6] based on the whole body CT images of a 24 kg female hound. The dog was selected considering the size and breed that can represent a broad range of canine subjects. Following the first canine model, they developed one more canine model based on the CT images of a 1.5 year old male Labrador [11] using a process almost identical to that used in the first canine model. The UF canine series was developed to establish a comprehensive library of photon and electron specific absorbed fraction (SAF) values for source and target organs. The data then were used to derive organ dose conversion coefficients, called *S* values, which convert accumulated activity of radionuclides in source organs to absorbed dose in target organs. After CT scans, the skeletons of the dogs were harvested for high-resolution microCT scans to develop detailed skeletal dosimetry models.

After anesthesia was induced intravenously, the dogs were scanned in a prone position using a multi-slice CT scanner. The x-ray tube potential and current were set at 120 kVp and 10 mA, respectively, with the CT slice thickness of 2 mm. The dogs were given contrast enhancement agents and scanned three times for head and neck, thorax and abdomen, and pelvis and tail. After the CT scan was completed, the dogs were euthanized, and the complete skeleton was harvested for *ex vivo* CT and *in vitro* microCT studies of the skeletal structure. Details about the skeletal model are not included in this chapter but can be found in the literature [6, 11].

The CT images were then imported to 3D-Doctor™, a 3D modeling and image processing software, for organ and tissue segmentation. Organs and tissues of interest were segmented manually or semi-automatically (possible for the lungs and skeleton) on transversal CT images slice by slice. A total of 29 major organs and 19 bone sites were segmented for model construction. The organs were selected based on radiosensitive organs listed in ICRP Publication 89 [12] reporting human radiological protection dosimetry data. Adipose tissue, skeletal muscle, blood vessels, lymphatic tissues, and connective tissues were merged and tagged as residual soft tissues (RST) without separate segmentations. Some smaller organs that were not visible from the CT images were manually created by referring to an anatomical textbook [13] and a practicing veterinary oncologist.

Once the segmentation was completed, polygon-mesh models representing organs and skeletons were exported from 3D-Doctor™ in the format called *Wavefront Object* (OBJ) files in five different groups: the external body contour, the alimentary system, the respiratory system, other soft-tissue organs, and the skeleton. The OBJ files were then imported into Rhinoceros™, 3D modeling software. To resolve the issue of rough surfaces mentioned in the previous section, NURBS modeling technique was used. Several sectional contours were generated from polygon mesh organ models and used to create smooth NURBS surfaces. The smoothing process was applied to all organs and tissues except for skeleton and outer body contour, which were too complex to model using NURBS surface. The final canine models in polygon mesh and NURBS surfaces were then converted into voxel models by using an in-house voxelization code [7] for Monte Carlo radiation transport simulations. Figures 6.1 and 6.2 show the UF Hound and Labrador models, respectively, in polygon mesh and NURBS format rendered by Rhinoceros™ software.

6.3.2 The NIRAS canine models

The researchers at The National Internal Radiation Assessment section's (NIRAS) Human Monitoring Laboratory (HML) had used different types of stylized and

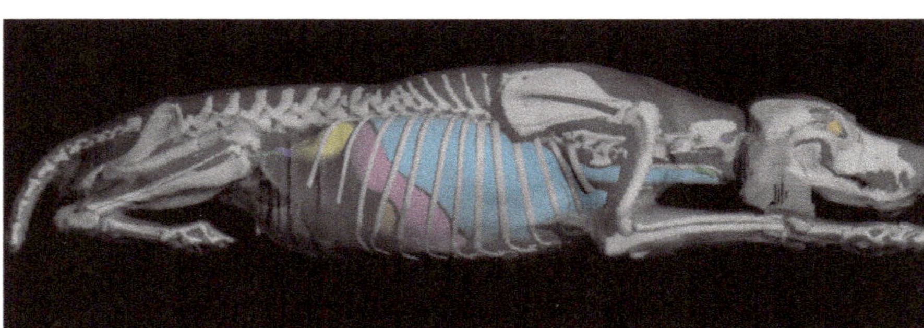

Figure 6.1. The UF Hound model in polygon mesh and NURBS format. The outer body contour was made semi-transparent to clearly visualize the skeleton and internal organs.

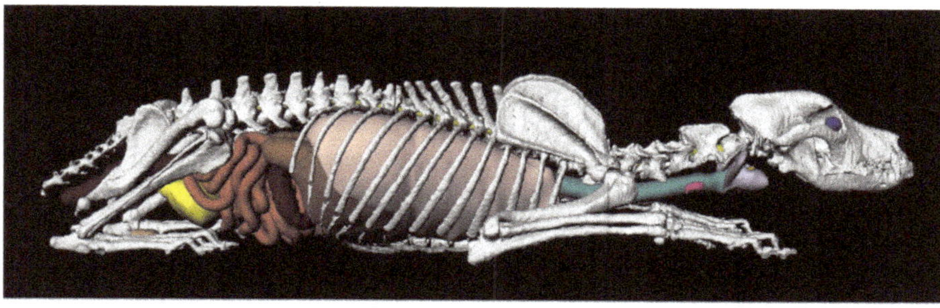

Figure 6.2. The UF Labrador model in polygon mesh and NURBS format. The outer body contour was made transparent to visualize the internal organs and skeleton.

voxel phantoms to help understand the effect of changing parameters on *in vivo* counting and the calibration of detector systems. They developed their own human and canine models for their research. This section describes the development of the two NIRAS canine models [14].

The digital imaging and communications in medicine (DICOM) CT images of a dog with 44 cm to withers and weight of 14 kg were obtained from Colorado State University. The same image set was used to create two canine models: NIRAS-K9 adolescent and adult models. The adolescent model was first created from the CT images and scaled up to create an adult size canine model. The dog was scanned in a supine position with the legs folded toward the body and the head horizontal with the spine so that the 3D modeling step was required to make the legs straight as shown in figure 6.3.

The CT images were imported into 3D-Doctor™, a computer program for image segmentation and 3D model generation. Major organs and tissues were manually segmented from the CT images. Within 3D-Doctor™, the lowered head was raised, and the folded legs were straightened manually.

Finally, the adolescent canine model was imported into a 3D modeling computer program, Rhinoceros™, to create an adult size canine model by scaling up the adolescent model based on the ratio of the height of the adolescent dog to that of the adult. The standard height of the adult canine model was obtained from literature.

6.3.3 The Vanderbilt University canine models

Stabin *et al* [15] reported two canine models based on the CT images of two adult beagles. The models were developed to calculate SAFs for internal organs and estimate dose conversion coefficients for selected radionuclides.

The two dogs were scanned in a Siemens CT scanner with the tube potential of 120 kVp and the tube current–time product of 47 mAs at the 5 mm slice thickness. Organs and tissues were segmented by using ITK-SNAP toolkit [16]. A total of 24 organs and tissues were included in the segmentation process. Complicated structures such as skeleton and intestines were modeled in the next step in Rhinoceros™ software by using polygon mesh and NURBS surfaces. Finally, the mesh models were converted to a voxel format for Monte Carlo radiation transport

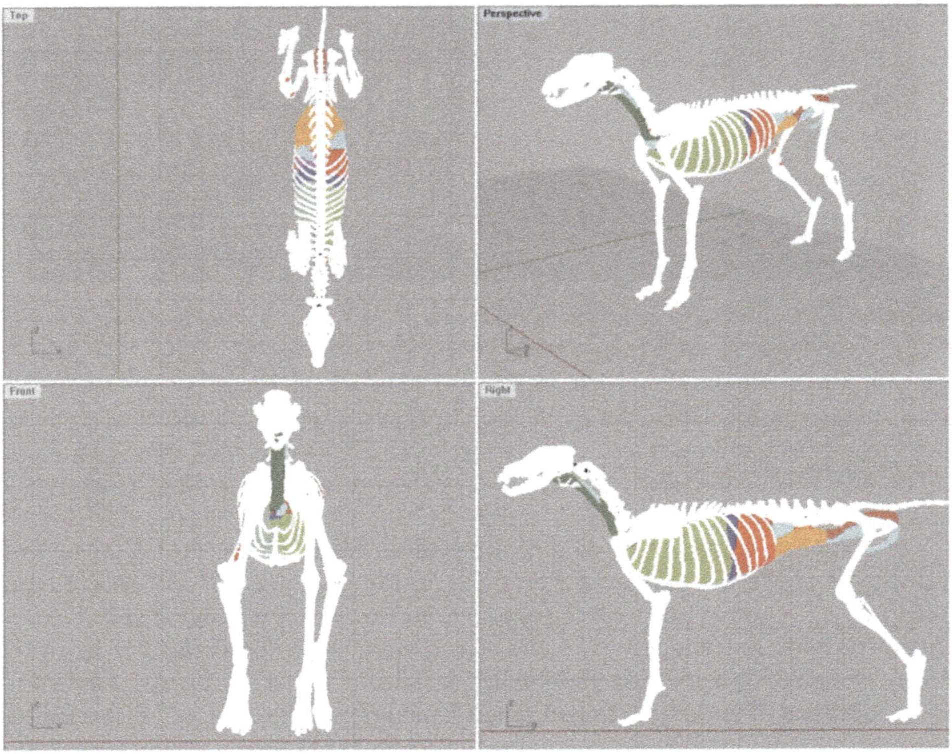

Figure 6.3. 3D rendering of the NIRAS-K9 voxel model shown in the *X*, *Y*, and *Z* planes and the perspective view. Reproduced with permission from Kramer *et al* [14].

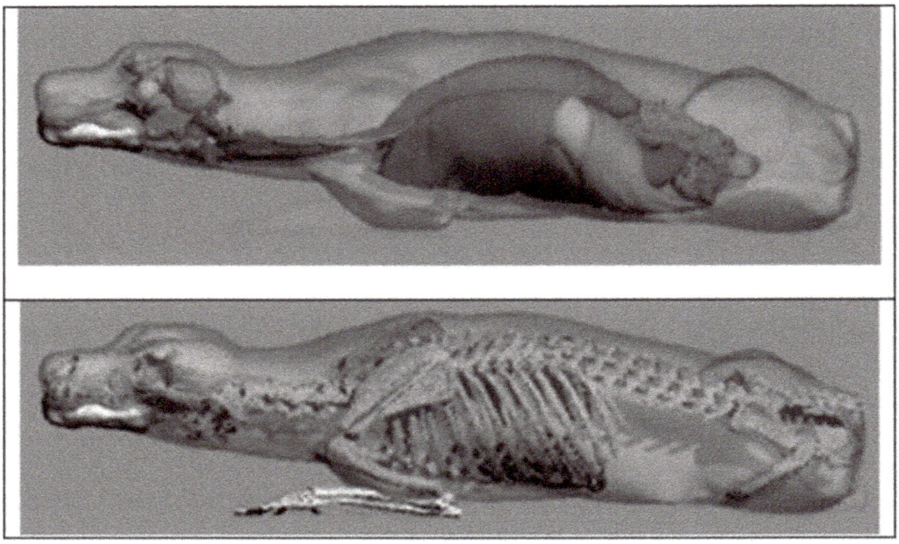

Figure 6.4. 3D rendering of the male adult canine model, Rusty, showing internal organs (top) and skeletons (bottom) with the outer body contour. Reproduced with permission from Stabin *et al* [15].

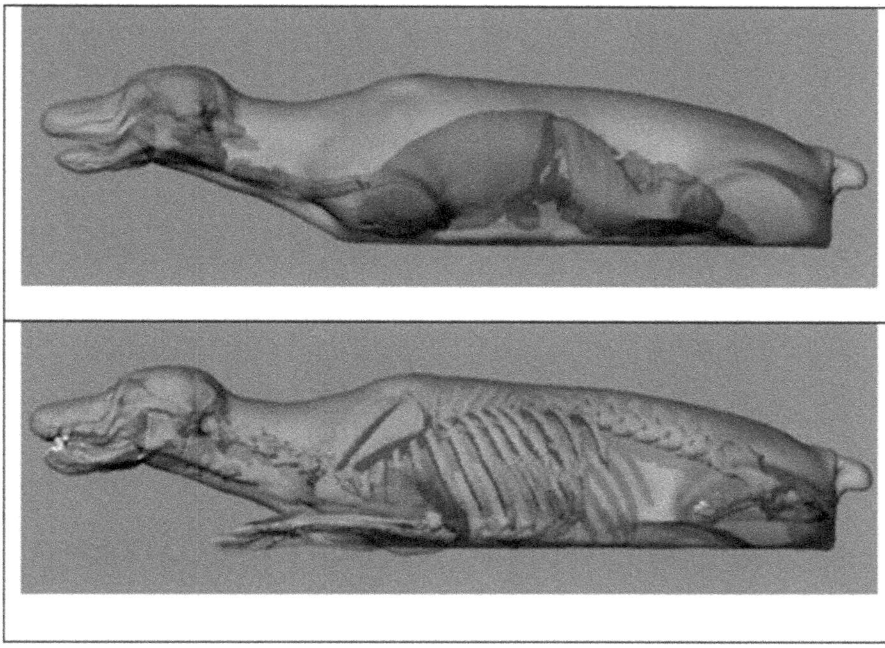

Figure 6.5. 3D rendering of the female adult canine model, Zena, showing internal organs (top) and skeletons (bottom) with the outer body contour. Reproduced with permission from Stabin *et al* [15].

calculations. Figures 6.4 and 6.5 show the 3D rendering of the male and female adult canine models developed in this study, respectively.

6.4 Summary and future perspectives

This chapter summarized the development process and characteristics of a total of six canine models that have been reported by different research groups to date. The major characteristics of the canine models mentioned in the previous sections are summarized in table 6.1 including age, gender, body size, breed, voxel resolution of the final voxel models, and purpose of the model development. The masses of the major organs (g) in the five canine models are summarized in table 6.2.

There has been significant progress in the development of computational human models since the introduction of the third-generation hybrid phantoms. More sophisticated tissues such as thin radiosensitive layer in walled organs [17] or detailed human eyes [18] have been reported to provide detailed and accurate dosimetry data. Polygon or tetrahedral mesh-based phantom technique has made this advanced modeling possible. These new developments are coupled with the direct Monte Carlo transport technique within polygon or tetrahedral mesh medium [19] to provide more accurate dosimetry results that were not available from the previous-generation computational phantoms. Small animal models can also take advantage of these new techniques to better describe small organ and tissue structures and conduct Monte Carlo radiation transport directly in surface-format models without the voxelization process which may sacrifice the details.

Table 6.1. Summary of the characteristics of the canine models reviewed in this chapter.

Canine models	Gender	Age (year)	Height (cm)	Weight (kg)	Breed	Voxel volume (cm^3)	Purpose	Reference
UF Hound	F	3	NA	24.0	Hound	0.008	Nuclear medicine	Padilla *et al* [6]
UF Labrador	M	1.5	NA	NA	Labrador	NA	Nuclear medicine	Sands *et al* [11]
Vanderbilt Rusty	M	Adult	NA	9.45	Beagle	0.001 125	Nuclear medicine	Stabin *et al* [15]
Vanderbilt Zena	F	Adult	NA	8.29	Beagle	0.001 125	Nuclear medicine	Stabin *et al* [15]
NIRAS-K9 (adolescent)	M	NA	44.0	14.0	Doberman	0.008 651	NA	Kramer *et al* [14]
NIRAS-K9 (adult)	M	NA	72.0	33.8	Doberman	0.005 222	NA	Kramer *et al* [14]

Table 6.2. Comparison of organ mass (g) of different canine models.

Organs	UF Hound	UF Labrador	Vanderbilt Rusty	Vanderbilt Zena	NIRAS-K9 Adult	NIRAS-K9 Adolescent
Adipose		6481.4			4768	1978
Adrenals	0.31	3.3				
Bladder	132.73[a]	28.0	28.9	13.9	129	54
Bone	2452.22	4493.0	971.60	791.50	5070	2098
Brain	86.15	115.0	65.12	65.25	88	36
Colon	496.13[a]	148.4	97.2[c]	118.00[c]	144	60
Eyes			11.90	10.41		
Esophagus	11.56	130.4[b]	17.20	13.43	27	11
Gall bladder			7.05	9.88		
Gonads		55.2	5.19	2.00		
Heart	424.13[a]	563.3[a]	199.00[e]	163.00[e]	566	234
Kidneys	202.77	243.4	80.40	79.80	134	56
Liver	707.80	1632.9	391.30	295.60	1000	418
Lungs	518.10	873.0	148.30	154.10	884	367
Muscle		19297.6			15978	6624
Pancreas	1.91	19.8	23.06	23.06		
Prostate			13.07			
Salivary glands			17.92	22.00		
Skin	2080.04	1047.2			2375	983
Small intestine	698.28	484.7	343.50	320.30	542	224
Spinal cord	18.83				53	22
Spleen	376.86	114.4	230.90	166.40	109	45
Stomach	397.55[a]	130.4[b]	151.80[d]	97.80[d]	1323	534
Thymus				2.71	2.57	
Thyroid	0.98	1.6		0.84	0.65	
Uterus					8.00	

[a] Mass of the wall and content were combined.
[b] Masses of the esophagus and stomach were combined.
[c] Only left large intestine is segmented and included in the mass.
[d] Only stomach contents are segmented and included in the mass.
[e] Only heart contents are segmented and included in the mass.

References

[1] Lindblad-Toh K, Wade C M, Mikkelsen T S, Karlsson E K, Jaffe D B and Kamal M *et al* 2005 Genome sequence, comparative analysis and haplotype structure of the domestic dog *Nature* **438** 803–19

[2] Starkey M P, Scase T J, Mellersh C S and Murphy S 2005 Dogs really are man's best friend--canine genomics has applications in veterinary and human medicine! *Brief Funct. Genomic Proteomic* **4** 112–28

[3] Vail D M and MacEwen E G 2000 Spontaneously occurring tumors of companion animals as models for human cancer *Cancer Invest.* **18** 781–92

[4] Xu X G 2014 An exponential growth of computational phantom research in radiation protection, imaging, and radiotherapy: A review of the fifty-year history *Phys. Med. Biol.* **59** R233–302

[5] Caon M 2004 Voxel-based computational models of real human anatomy: A review *Radiat. Environ. Biophys.* **42** 229–35

[6] Padilla L, Lee C, Milner R, Shahlaee A and Bolch W E 2008 A canine anatomical phantom for preclinical dosimetry in molecular radiotherapy *J. Nucl. Med.* **49** 446–52

[7] Lee C, Lodwick D, Hurtado J, Pafundi D, Williams J L and Bolch W E 2010 The UF family of reference hybrid phantoms for computational radiation dosimetry *Phys. Med. Biol.* **55** 339–63

[8] Goorley T, James M, Booth T, Brown F, Bull J and Cox L J *et al* 2012 Initial Mcnp6 Release Overview *Nucl. Technol.* **180** 298–315

[9] Allison J, Amako K, Apostolakis J, Arce P, Asai M and Aso T *et al* 2016 Recent developments in Geant4 *Nucl. Instrum. Methods Phys. Res. Section* A **835** 186–225

[10] Sato T, Iwamoto Y, Hashimoto S, Ogawa T, Furuta T and Abe S-i *et al* 2018 Features of Particle and Heavy Ion Transport code System (PHITS) version 3.02 *J. Nucl. Sci. Technol.* 1–7

[11] Sands M M 2016 Hybrid computational phantom of the labrador with a detailed skeletal model for radiopharmaceutical therapy dosimetry *Ms Thesis* (University of Florida)

[12] ICRP. Basic anatomical and physiological data for use in radiological protection: reference values. Oxford; Pergamon Press: International Commission on Radiological Protection; 2002 September 2001. Report No.: Publication 89

[13] Miller M E 1993 *Anatomy of the Dog* (Philadelphia, PA: WB Saunders Company)

[14] Kramer G H, Capello K, Strocchi S, Bearrs B, Leung K and Martinez N 2012 The HML's new voxel phantoms: Two human males, one human female, and two male canines *Health Phys.* **103** 802–7

[15] Stabin M G, Kost S D, Segars W P and Guilmette R A 2015 Two realistic beagle models for dose assessment *Health Phys.* **109** 198–204

[16] Yushkevich P A, Piven J, Hazlett H C, Smith R G, Ho S and Gee J C *et al* 2006 User-guided 3D active contour segmentation of anatomical structures: Significantly improved efficiency and reliability *Neuroimage* **31** 1116–28

[17] Yeom Y S, Kim H S, Nguyen T T, Choi C, Han M C and Kim C H *et al* 2016 New small-intestine modeling method for surface-based computational human phantoms *J. Radiol. Prot.* **36** 230–45

[18] Nguyen T T, Yeom Y S, Kim H S, Wang Z J, Han M C and Kim C H *et al* 2015 Incorporation of detailed eye model into polygon-mesh versions of ICRP-110 reference phantoms *Phys. Med. Biol.* **60** 8695–707

[19] Yeom Y S, Jeong J H, Han M C and Kim C H 2014 Tetrahedral-mesh-based computational human phantom for fast Monte Carlo dose calculations *Phys. Med. Biol.* **59** 3173–85

Chapter 7

Overview of computational rabbit models

Kanako Wake, Akimasa Hirata and Kenji Taguchi

7.1 Introduction

For non-uniform or localized non-ionizing radiation, adverse/sensory effects in the eye are considered as the rationale for international safety guidelines/standards. For low-frequency exposures (lower than 300 Hz), the phosphene caused by the *in situ* electric field in the retina [1] is considered as a basis because the corresponding external field strength is, among others, one of the lowest. For microwave exposures (approximately GHz bands), cataract formation, caused by excess temperature elevation in the lens for a certain duration, is considered as the rationale for human protection [2]. For the latter, it is impossible to conduct volunteer studies and so animal studies are often conducted.

Rabbits are often used because their eye structure is relatively similar to that of humans. In an earlier study [2, 3], cataract formation was reported in the rabbit eye exposed to microwaves at 2.45 GHz. Thereafter, several studies have been reported on cataract formation in the eye (e.g. [4, 5]). Thus, to investigate the rationale of the international safety guidelines [6, 7], computational rabbit models are needed. Particular attention should be paid to the eye modeling.

Historically, the modeling of the eye was much simplified as an object thermally isolated from the head [8, 9]. In the simplification, the thermal flow at the boundary between the eyeball and body core was represented in terms of a convection coefficient. At that time, it was uncertain if high blood perfusion rate in the choroid and retina was well considered. Later, the importance of such tissues was discussed [10]. For small animals, including rabbits, the temperature in other tissues including the body core may elevate, and thus modeling in other tissues is not negligible [2, 11]. Despite the importance of the rabbit eye in non-ionizing radiation dosimetry, only one numerical rabbit model developed by the present authors is available.

In this chapter, we review its initial development as well as recent progress in model refinement.

doi:10.1088/2053-2563/aae1b4ch7

7.2 Construction of rabbit models

7.2.1 Acquisition of CT images

The numerical model was developed from x-ray CT images of an anesthetized rabbit constrained by a plastic holder made of polycarbonate, taken at Kanazawa Medical University [12]. The plastic holder is specially designed for immobilization to acquire the images and to conduct animal exposure studies.

The CT images were taken at 1 mm intervals in the anteroposterior direction of the rabbit with a body length of 325 mm. The resolution of each image was 2.13 pixels/mm. The CT images were recorded in digital imaging and communication in medicine (DICOM) format with 16 bit gray-scale. A photograph of the rabbit in the plastic holder and a representative CT image are illustrated in figure 7.1.

7.2.2 Tissue classification from CT images

The conversion from CT images to a numerical (voxel) model was partly automated using a segmentation method similar to the one described in [13]. To each pixel is assigned a CT value (Hounsfield unit) reflecting the absorption rate of x-rays relative to water. The CT value of water is 0 and that of air as low as −1000 since it doesn't absorb x-rays. The CT value of bone and fat is known to be around 1000 and −100, respectively, However the CT value of soft tissue such as the brain, muscles, etc is around 100 and, as such, it is difficult to distinguish them from each other. Therefore, we automatically segmented each pixel to different tissue classes by considering CT values and knowledge of body structures. We modeled the rabbit eye in detail by manual segmentation as it was insufficient to model the eye from CT images with the above-mentioned resolution. The outcome of manual segmentation was checked by an ophthalmologist. The developed model consisted of 12 tissue types: skin, muscle, bone, fat, brain, CSF, aqueous, vitreous, retina/choroid/sclera, iris/ciliary body, lens, and cornea. One limitation is that the thickness of most eye tissues, such as the retina, choroid, and so forth are smaller than the spatial resolution of the model. Therefore, the retina, choroid and sclera are considered as a mixed tissue and as such, their average physical (electrical and thermal) constants

(a) (b)

Figure 7.1. (a) Rabbit in the plastic holder and (b) an example of a CT image taken of the immobilized rabbit.

Table 7.1. Summary of tissue classification algorithm from CT images.

Tissues	CT value	Protocol
Air (anterior)	−2048 to −478	Automatic segmentation using CT values
Air (Interior)	−2048 to −478	Automatic segmentation using CT values
Bone	178 to 2570	Automatic segmentation using CT values
Fat	−295 to −4	Automatic segmentation using CT values
Muscle	Others	Automatic segmentation using CT values
Skin	Others	Automatic segmentation as tissue contacting anterior air (known thickness—specified in the code)
Brain	Others	Semi-manual segmentation
CSF	Others	Automatic segmentation as surroundings of the brain (known thickness—specified in the code)
Vitreous	Others	Manual segmentation
Retina/choroid/sclera	Others	Manual segmentation as surroundings of vitreous (not contacting anterior air)
Cornea	Others	Automatic segmentation as surroundings of vitreous (contacting anterior air)

were assigned in dosimetry calculations [14]. Table 7.1 summarizes the segmentation protocol for each tissue. The original computational rabbit model with a resolution of 1 mm is illustrated in figure 7.2.

7.3 Model refinement

Currently, a novel algorithm has been developed by different groups to enable refinement of the model, allowing the generation of a model with arbitrary resolutions [15]. Except for a few models based on computer-aided design, the resolution of the computational human/animal models is 1–3 mm, which is difficult to change. This was one of the significant obstacles encountered when conducting detailed dosimetry calculations, including dosimetry at higher frequencies where the wavelength in biological tissues is comparable to the model resolution. We then proposed a method for fine-tuning of the conventional voxel models. The method is composed of the following three steps: (1) polygonization of the conventional voxel model, (2) smoothing of the polygon model, and (3) voxelization of the smoothing polygon model. The smoothing and arbitrary resolution of the voxel model can be realized by mediating a polygon model. The schematic explanation of the whole process is illustrated in figure 7.3. The eye in the rabbit model with different resolutions is shown in figure 7.4.

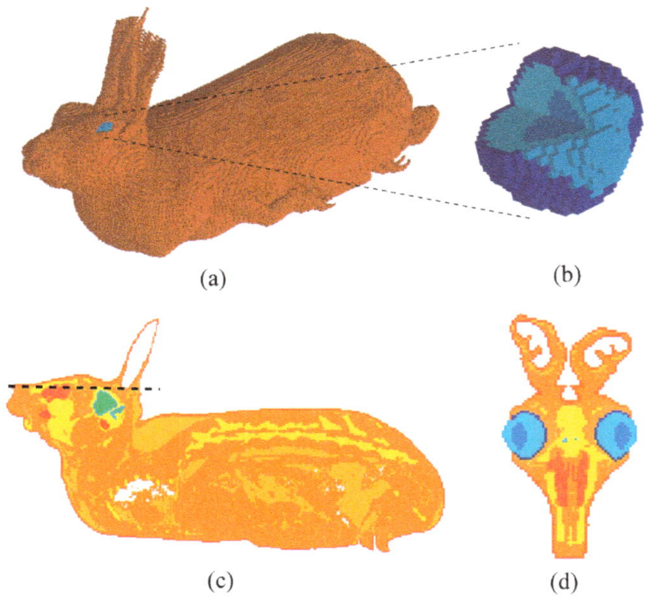

Figure 7.2. (a) Birds-eye-view of the rabbit model (original resolution of 1 mm) and (b) 3D view of the eyeball. Cross-section views of the rabbit model: (c) around the model center and (d) around the eye.

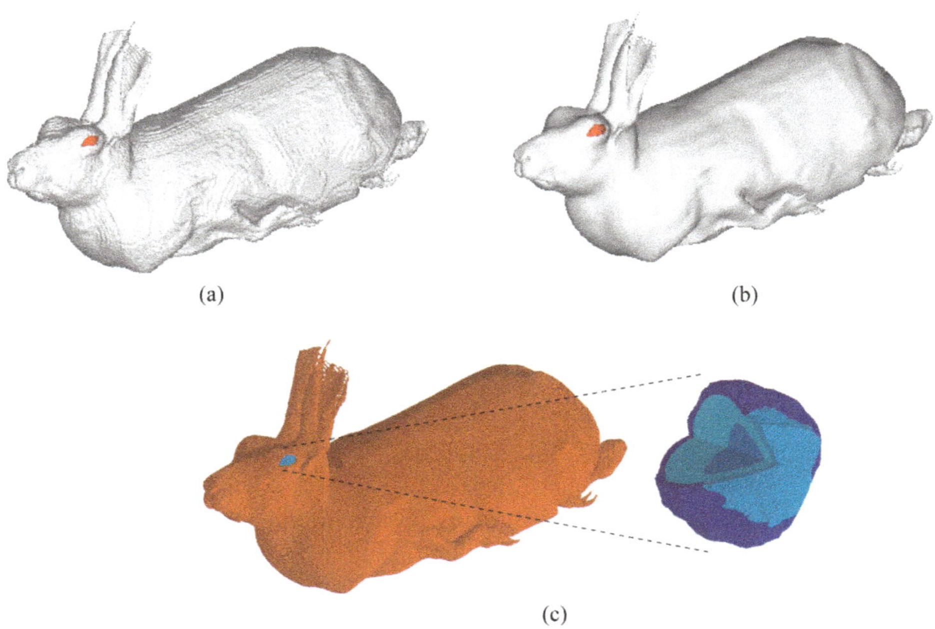

Figure 7.3. (a) Polygonized rabbit model from the original model (shown in figure 7.2) and its (b) smoothed model. (c) Voxelized model with a resolution of 0.125 mm.

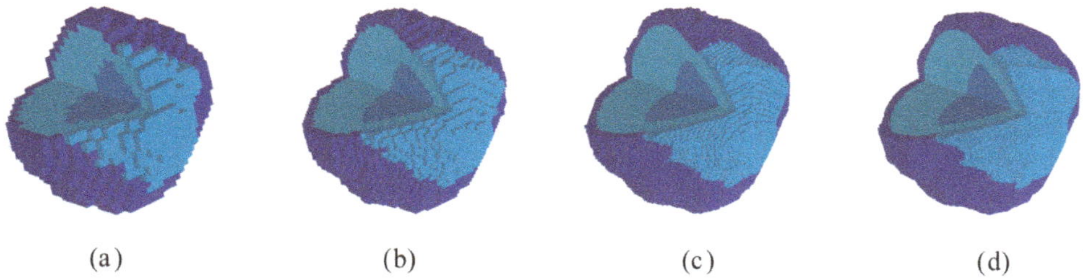

Figure 7.4. Eye extracted from the rabbit model with different resolutions: (a) 1.0 mm, (b) 0.5 mm, (c) 0.25 mm, and (d) 0.125 mm.

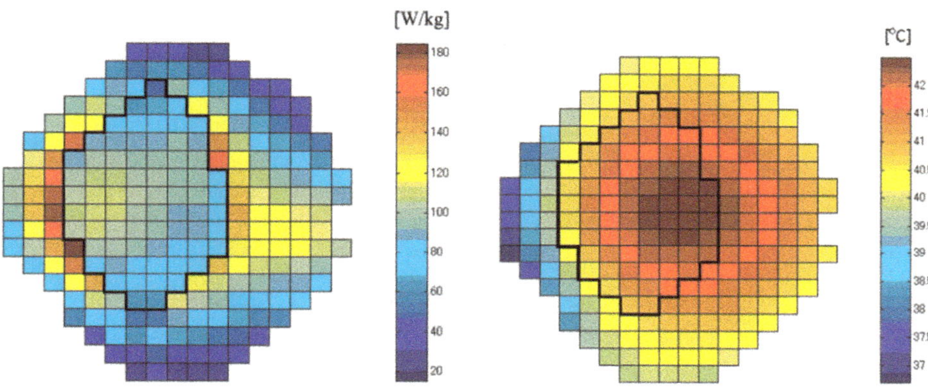

Figure 7.5. (a) SAR and (b) temperature distributions (at 40 min) on a horizontal cross-section of the rabbit eye under 2.45 GHz microwave exposure at a power density of 300 mW cm^{-2}. Reproduced from [14].

7.4 Examples of electromagnetic and thermal dosimetry

To demonstrate the effectiveness of our model, electromagnetic and thermal dosimetry was conducted for microwave exposure at 2.45 GHz. For localized exposure of the eye, a dielectric-filled waveguide antenna developed in [12] was used. The aperture dimension of the waveguide was 43.2 mm × 86.4 mm. The ratio of the eye averaged specific absorption rate (SAR—power deposition per unit mass) to the whole-body averaged SAR is 70. During microwave exposure, temperatures of the eye segments were measured with a Fluoroptic thermometer. The rabbit eye was anesthetized with 0.4% oxybuprocaine hydrochloride ophthalmic solution applied as eye drops.

Figure 7.5 shows the calculated SAR and temperature distribution at 40 min on the horizontal cross-section of the exposed eye. From figure 7.5(a), high SAR values are observed in the aqueous and vitreous, which have high conductivities. A comparison between figures 7.5(a) and (b) shows a clear difference in the distributions between SAR and temperature. There are two main reasons for this difference: the first is heat diffusion, whereas the second is the initial temperature distribution. As can be seen from these results, the assignment of inhomogeneous electrical conductivity and blood flow in eye tissues is essential for dosimetry calculations, which demonstrates the necessity of a realistic eye model.

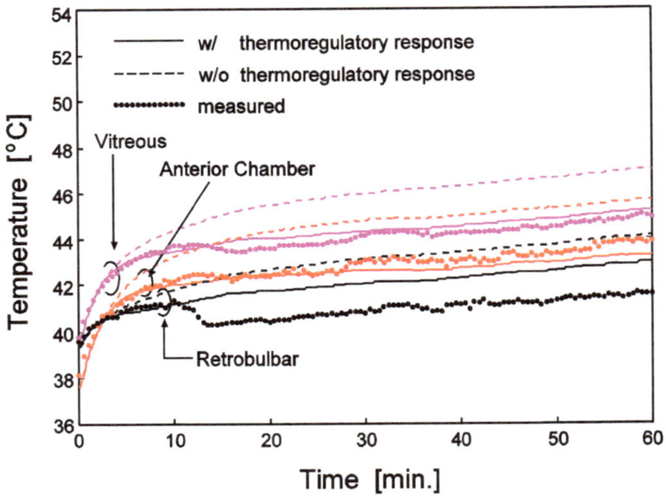

Figure 7.6. Measured and computed temperature in the rabbit eye without anesthesia (300 mW cm^{-2}). The rabbit eye was first irradiated to 2.45 GHz microwave for 60 min without anesthesia. The temperature when disregarding thermoregulatory response is also drawn for comparison. Reproduced from [14]. Copyright © 2006 Wiley-Liss, Inc.

Next, let us validate the computational model using measured data [5]. First, fifteen minutes after the temperature probes were inserted into the eye (after the anesthesia had worn off), the eyes were exposed to 300 mW cm^{-2} for 1 h, followed by a cooling time of 1 h. During that time, the temperature reached a thermally steady state. Fifteen minutes after the administration of systemic anesthesia, the rabbit was again exposed to 300 mW cm^{-2} for 1 h.

Calculated temperature variations in the rabbit eye are illustrated in figure 7.6 for cases with local anesthesia to the eye. As shown in figure 7.6(a), computed temperatures are shown for cases with and without thermoregulatory response. The temperatures with a thermoregulatory response were in good agreement with the measured data without anesthesia. The computed results without thermoregulatory response are overestimated compared with the measurements. The comparison suggests the necessity of the thermoregulatory response especially in the retina/choroid whose blood perfusion is high, again suggesting the necessity of a high resolution model around the eye of the rabbit.

Radio-frequency dosimetry with a model with higher resolution becomes more important because the new wireless technology called fifth generation mobile systems will be deployed soon. The frequency used in the system may include millimeter waves. Higher resolution models of the rabbit may then be required as shown in figures 7.3 and 7.4.

7.5 Summary

In this chapter, a detailed procedure for developing computational rabbit models has been presented. Special attention has been paid to the eye structure, because this organ is important for non-ionizing radiation protection.

Computational whole-body rabbit models may be used for complex radio-wave environments. For frequencies higher than millimeter waves, including laser radiation, a detailed modeling of the eye structure becomes more important for investigating the thermal damage as well as local thermoregulation.

References

[1] Lövsund P, Öberg P, Nilsson S and Reuter T 1980 Magnetophosphenes: a quantitative analysis of thresholds *Med. Biol. Eng. Comput.* **18** 326–34

[2] Guy A W, Lin J C, Kramar P O and Emery A F 1975 Effect of 2450-Mhz radiation on the rabbit eye *IEEE Trans. Microw. Theory Tech.* **23** 492–8

[3] Emery A, Kramar P, Guy A and Lin J 1975 Microwave induced temperature rises in rabbit eyes in cataract research *J. Heat Transfer* **97** 123–8

[4] Kamimura Y, Saito K-i, Saiga T and Amemiya Y 1994 Effect of 2.45 GHz microwave irradiation on monkey eyes *IEICE Trans. Commun.* **77** 762–5

[5] Kojima M, Hata I, Wake K, Watanabe S, Yamanaka Y and Kamimura Y *et al* 2004 Influence of anesthesia on ocular effects and temperature in rabbit eyes exposed to microwaves *Bioelectromagnetics* **25** 228–33

[6] ICNIRP 1998 Guidelines for limiting exposure to time-varying electric, magnetic, and electromagnetic fields (up to 300 GHz) *Health Phys.* **74** 494–521

[7] IEEE-C95.1 2005 *IEEE standard for safety levels with respect to human exposure to radio frequency electromagnetic fields, 3 kHz to 300 GHz* (NY, USA: IEEE)

[8] Lagendijk J 1982 A mathematical model to calculate temperature distributions in human and rabbit eyes during hyperthermic treatment *Phys. Med. Biol.* **27** 1301

[9] Scott J A 1988 A finite element model of heat transport in the human eye *Phys. Med. Biol.* **33** 227

[10] Hirata A 2007 Improved heat transfer modeling of the eye for electromagnetic wave exposures *IEEE Trans. Biomed. Eng.* **54** 959–61

[11] Hirata A, Kojima M, Kawai H, Yamashiro Y, Watanabe S and Sasaki H *et al* 2010 Acute dosimetry and estimation of threshold-inducing behavioral signs of thermal stress in rabbits at 2.45-GHz microwave exposure *IEEE Trans. Biomed. Eng.* **57** 1234–42

[12] Wake K, Hongo H, Watanabe S, Taki M, Kamimura Y and Yamanaka Y *et al* 2007 Development of a 2.45-GHz local exposure system for *in vivo* study on ocular effects *IEEE Trans. Microw. Theory Tech.* **55** 588–96

[13] Chou C, Chan K, McDougall J and Guy A 1999 Development of a rat head exposure system for simulating human exposure to RF fields from handheld wireless telephones *Bioelectromagnetics* **20** 75–92

[14] Hirata A, Watanabe S, Kojima M, Hata I, Wake K and Taki M *et al* 2006 Computational verification of anesthesia effect on temperature variations in rabbit eyes exposed to 2.45 GHz microwave energy *Bioelectromagnetics* **27** 602–12

[15] Taguchi K, Kashiwa T and Hirata A 2018 Development on high resolution human voxel model for high frequency exposure analysis *Prog. Electromagn. Res.*

Chapter 8

Overview of other computational animal models

Tianwu Xie and Habib Zaidi

8.1 Introduction

Exposure to ionizing radiation may result in devastating health effects to humans and other species. Therefore, the preparedness of radiation and radiological countermeasures are important security topics for radiation protection of individuals (staff and patients) and the environment. The hazard of ionizing radiation on organisms can be categorized into direct damage and indirect damage according to the biological effects and interacting mechanisms. Direct damage occurs when high energy radiation interacts with biological tissues resulting in functional and structural alterations of molecules, thus causing direct lesions in cells. The reported direct DNA-damage in cells from ionizing radiation includes different base damage, single-strand breaks, double-strand breaks and DNA cross links [1]. Alternatively, indirect damage from radiation is caused by radiation-induced reactive and toxic free species and radicals, which lead to several pro-inflammatory reactions and tissue injuries [2]. These toxic species and radicals are produced by the interaction of intracellular or extracellular water and ionizing radiation. The combination of direct damage and indirect damage leads to different types of radiation injuries depending on the absorbed radiation dose in tissues and the dose distribution in the body. The extent of radiation-induced injury on organisms can be estimated by measuring the biodosimetry and physical dosimetry to the gastrointestinal system, neurovascular system, lymphohematopoietic tissues, skin and lungs. Different radiation countermeasures have been developed to protect individuals from radiation injury and reducing the risks of radiation lethality [3]. Laboratory animals are widely used in the development of radiation countermeasures to evaluate the harmful effects of radiation exposure on similar human tissues and organs [4]. Rodents are outstanding small animal models for the estimation of hematopoietic injuries, gastrointestinal injuries and lung injuries as they can be easily and economically bred in laboratories. However, the disadvantage of rodents is that their organ structure significantly

doi:10.1088/2053-2563/aae1b4ch8

differs from humans and they have accelerated life spans and small body weights. As well as rodents, commonly used animals for radiation dosimetry include pigs, rabbits, ferrets, canines and non-human primates, etc. Computational animal models were developed to mimic the physical properties and anatomy of laboratory animals to simulate radiation transport of ionizing particles in organisms and provide estimations of absorbed dose and dose distributions in laboratory animals to develop radiobiological relationships between radiation exposure parameters (e.g. absorbed dose rate and radiation quality, etc) and radiation injuries. For environmental radiation protection, in order to assess the relationships between ionizing radiation exposures and the dose effects on biota, a set of reference animals were developed and used to calculate the absorbed doses in relevant organs [5].

The computational models of rats, mice, frogs, canines and rabbits have been summarized in previous chapters. In this chapter, we will review the published computational models of other animals, including trout, crabs, flatfish, bees, deer, earthworms, ducks, goats, pigs and non-human primates, and their applications in radiation dosimetry research.

8.2 Computational models of trout

Trout live in freshwater throughout the world and are considered as a biological indicator of water quality in different environmental legislations. They are used in many laboratories for evaluating the radiation effects of radionuclides in marine biota. To protect the marine environment and calculate the radiation dose to trout from radioactive environmental contaminants, the ICRP [6] constructed a stylized model of the trout and reported the corresponding dose conversion factors in external exposures from 75 radionuclides. Hess *et al* [8] constructed a voxel model of rainbow trout based on CT and MR images and reported the absorbed organ dose of monoenergetic electrons and photons in the trout. Figure 8.1(b) shows the construction of the rainbow trout model by segmenting internal organs in CT images. Martinez *et al* [9, 10] constructed a stylized model, a voxel model and a hybrid model of rainbow trout based on CT images. Figure 8.2 shows the developed three types of models by Martinez *et al* [9, 10].

8.3 Computational models of crabs

Crabs have served as representative crustaceans for environmental radiation protection and were also used in laboratories for radiochemical and radiobiological research. The ICRP [6] constructed a stylized reference crab model of the Cancrid Super-Family crab and calculated dose conversion factors. Caffrey *et al* [7] developed a voxel model of a 464 g Dungeness crab and reported the absorbed dose of electrons and photons in the model. Figure 8.3 shows the segmentation process based on CT images and the final developed voxel-based crab model.

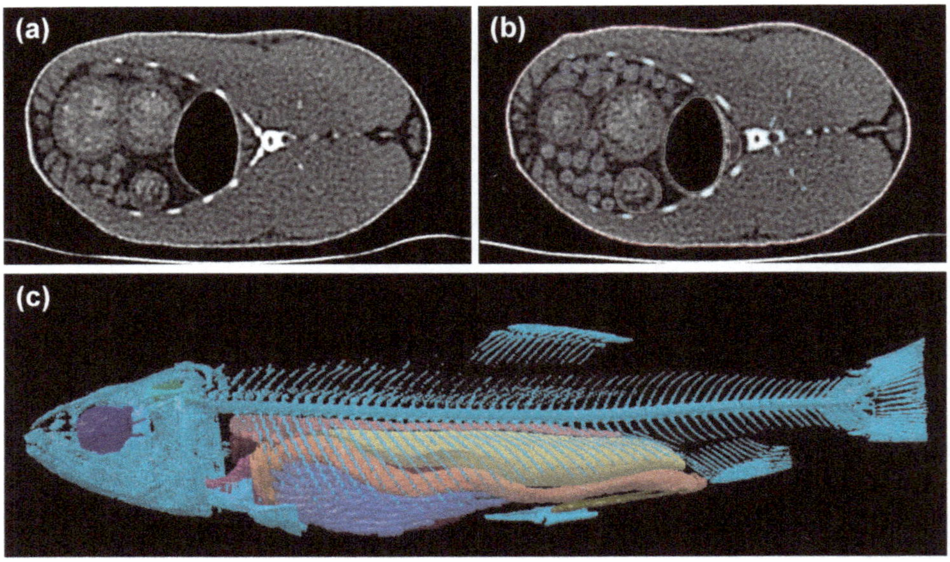

Figure 8.1. Construction of voxel-based trout model by Hess *et al* (a) Representative CT image; (b) segmentation of internal organs; (c) constructed trout model. (Reproduced with permission from reference [8].)

8.4 Computational models of flatfish

Flatfish live in both brackish waters and marine waters. It is an important commercial marine fish in many countries and has been used in laboratories to investigate the accumulation of radionuclides in fish and the potential radiation effects. The ICRP [6] published a stylized reference flatfish model of the Pleuronectid family and reported dose conversion factors. Caffrey *et al* [11] of Oregon State University constructed a voxel model of a 1024 g Pleuronectid flatfish based on CT and MR images for radiation dosimetry research. Figure 8.4 shows CT and MR images used for the construction of voxel-based flatfish models.

8.5 Computational models of bees

The bee is a typical insect in a terrestrial environment in many parts of the world, and its population has been used as a biological indicator for the environment or surrounding climate. The ICRP [6] used a stylized reference bee model of the typical Family Apidea bee for radiation dosimetry in the context of environmental radiation protection. Gomez *et al* [12] constructed a voxel-based model of the *Apis mellifera* honeybee using CT images and calculated the absorbed fraction of electrons and photons in their model. Figure 8.5 shows the CT image of a honey bee organism from a top view angle and the 3D visualization of the developed voxel model.

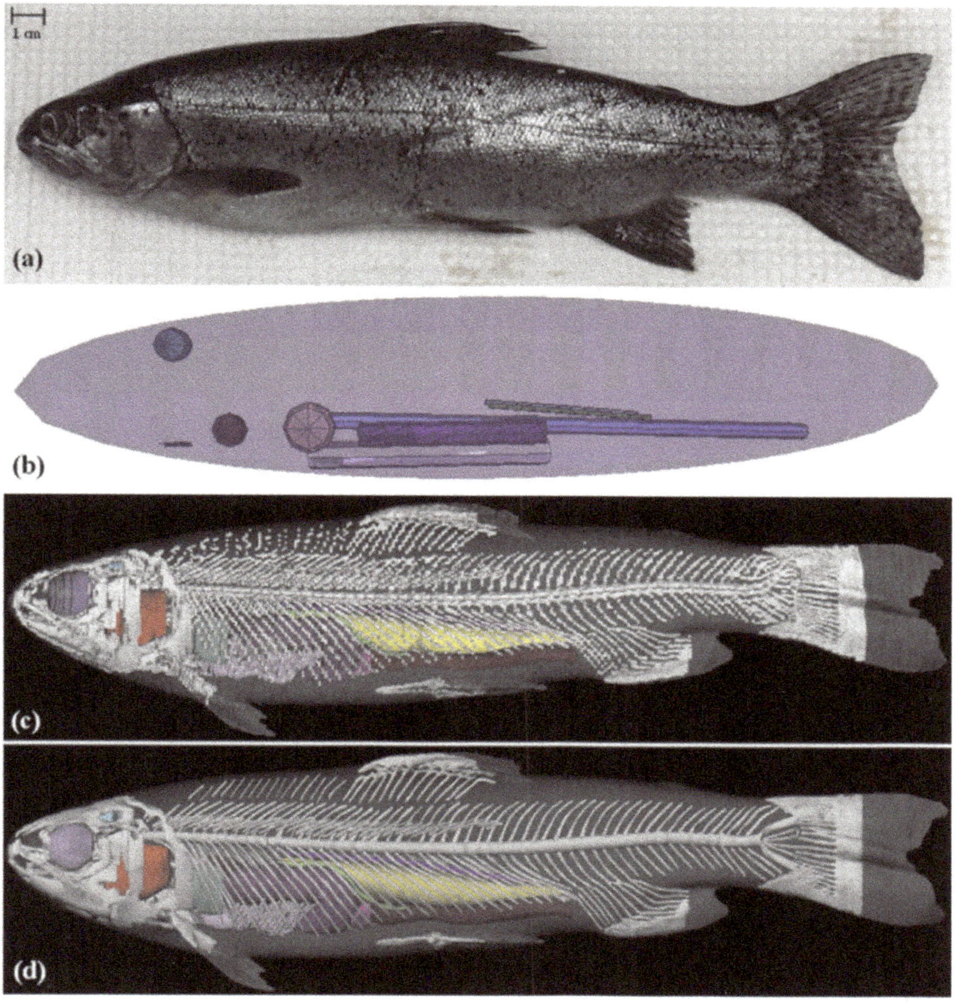

Figure 8.2. The trout models developed by Martinez *et al* (a) The selected organism; (b) stylized trout model; (c) voxel trout model and (d) the hybrid trout model. (Reproduced with permission from references [9, 10]. Copyright Elsevier.)

8.6 Computational models of deer

Deer are large terrestrial mammals which occur throughout many countries all over the world. The ICRP [6] reported a stylized reference deer model of a medium-sized woodland deer and calculated the radiation dose rate for environmental radiation protection. Figure 8.6 shows the stylized deer model with the body, liver and testes represented by ellipsoids of different sizes.

8.7 Computational models of earthworms

Earthworms play a vital part within the natural soil ecosystem for the decomposition of organic materials and changing soil physical properties. They also are an

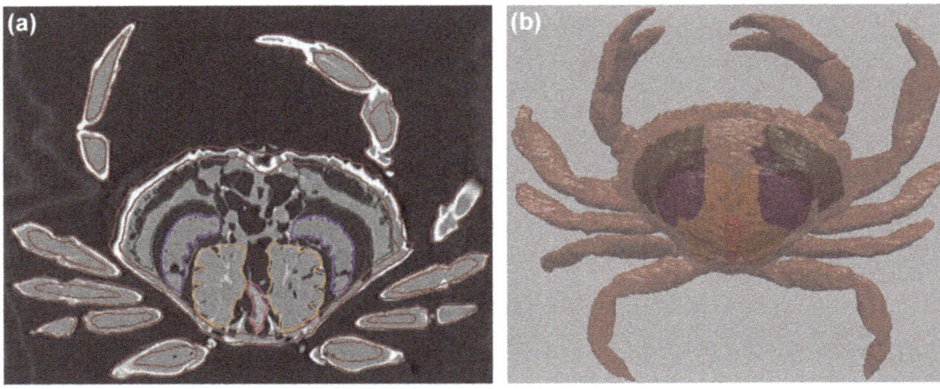

Figure 8.3. Construction of the crab model by Caffrey *et al.* (a) The segmentation of internal organs based on CT images of the selected crab; (b) the developed voxel model of the crab. (Reproduced with permission from reference [7]. Copyright 2013, with permission from Elsevier.)

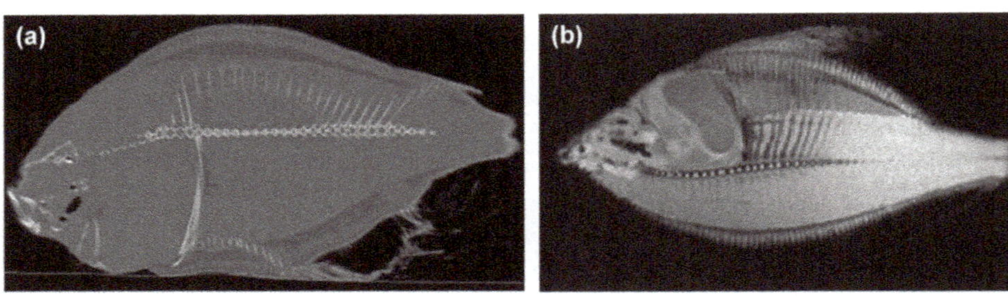

Figure 8.4. (a) The CT images and (b) MR images used by Caffrey *et al* for constructing the voxel-based flatfish model. (Reproduced with permission from reference [11]. Copyright by Emily Amanda Caffrey.)

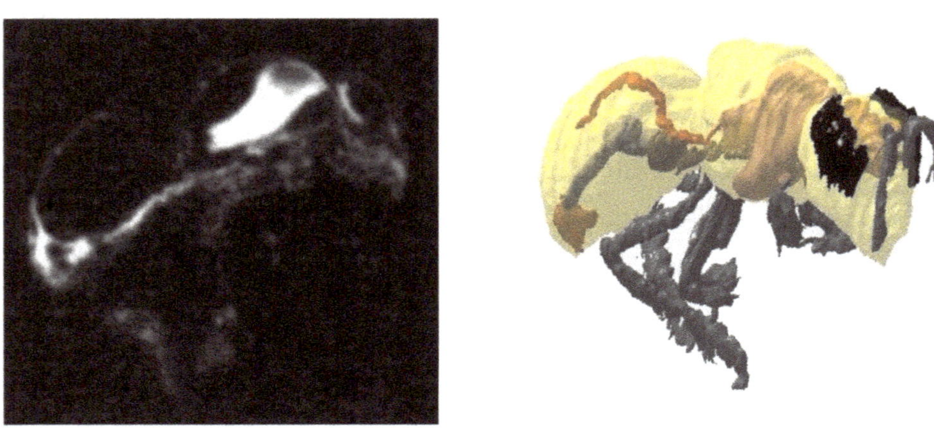

Figure 8.5. (a) CT image of a honeybee organism from a top view angle and (b) the 3D visualization of the developed voxel model of the honeybee. (Reproduced with permission from reference [12]. Copyright by Mario Enrique Gomez Fernandez, 2016. All Rights Reserved.)

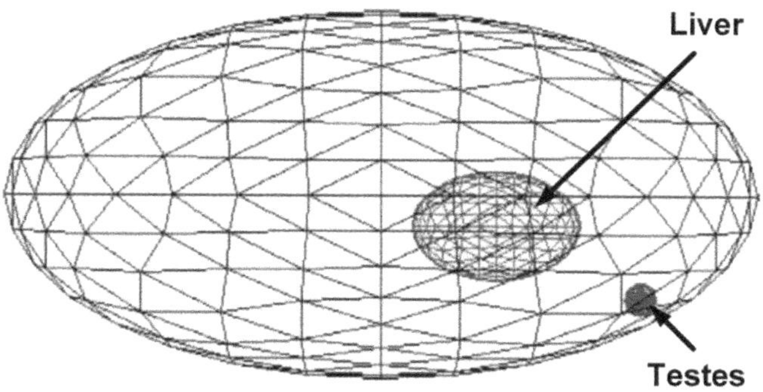

Figure 8.6. Geometrical model of a reference deer developed by the ICRP. (Reproduced with permission from reference [6]. Copyright © 2018 by SAGE Publications.)

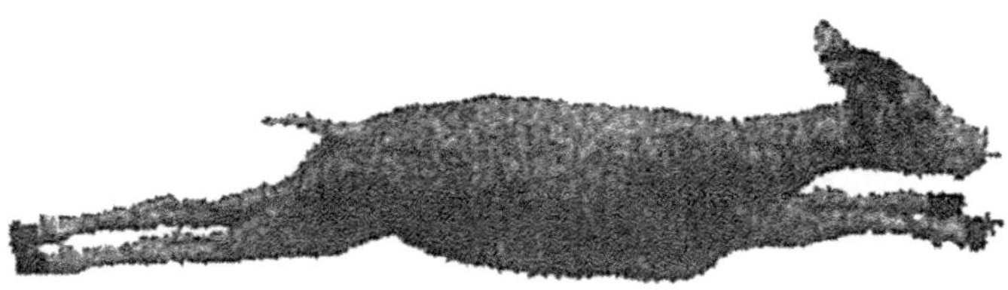

Figure 8.7. The goat model developed by Mason *et al* (Reproduced with permission from reference [13]. © Springer Science+Business Media New York 1999.)

important food source for many mammals and birds. They can live in different types of agricultural soils and have been used as indicators of environmental contamination. The ICRP [6] constructed a stylized reference earthworm model of the Lumbricid Family for environmental radiation protection.

8.8 Computational models of ducks

Domestic ducks and wild ducks live in both urban and rural areas in many parts of the world and are commonly used as food source for humans in various countries. They were chosen as typical organisms in terrestrial and aquatic environments. The ICRP [6] published a stylized reference duck model for radiation dose estimation from radionuclides.

8.9 Computational models of goats

Goats are one of the most ubiquitous medium-sized land mammals which can be found all over the world. They have been used as an established model in biomedical research in many laboratories. Mason *et al* [13] from Systems Research Laboratories of Brooks Air Force Base developed a voxel model of a 20 kg pigmy goat based on MR images and calculated the specific absorption rate from implanted temperature probes, heating-sensitive paints and infrared imaging. Figure 8.7 shows the goat model developed by Mason *et al* [13].

8.10 Computational models of pigs

The pig is a very good animal model for investigating lesions of the skin following radiation exposure as the physiology of the skin is similar to humans. Toivonen *et al* [14] from Radiation and Nuclear Safety Authority of Finland constructed a voxel model of a 25 kg pig based on CT and MR images. They calculated and reported the radiofrequency electromagnetic dosimetry for pigs from 900 MHz GSM mobile phone fields.

8.11 Computational models of non-human primates

The non-human primate (such as rhesus macacus and *Macaca fascicularis*) is considered to be the gold standard animal model enabling radiation injury of the human body to be mimicked because its organ structure, genome, life span and metabolism are close to those of humans. However, the breeding, housing and feeding of non-human primates costs much more than other laboratory animals and there are ethical considerations around research on non-human primates [4]. Mason *et al* [13] from Systems Research Laboratories of Brooks Air Force Base in USA constructed two voxel models of a rhesus monkey and a phantom monkey based on MR images. They calculated specific absorption rate values in the non-human primate from implanted temperature probes, heating-sensitive paints and infrared imaging. Figure 8.8 shows the constructed rhesus monkey model and phantom monkey model by Mason *et al* [13].

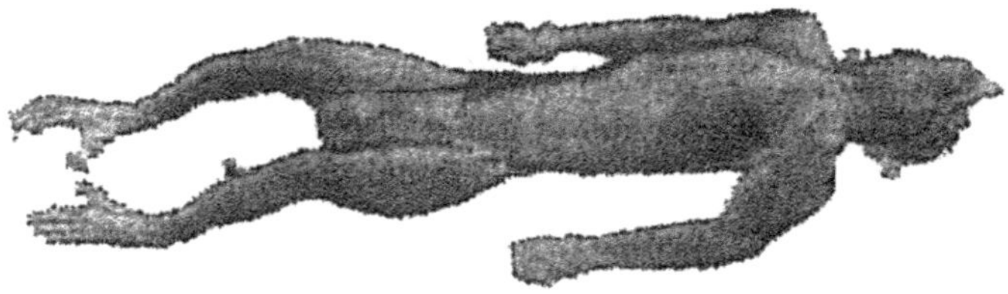

(a)

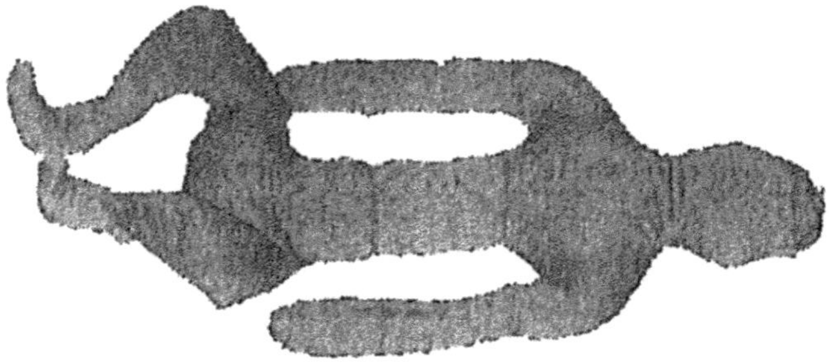

(b)

Figure 8.8. (a) The rhesus monkey model and (b) phantom monkey model developed by Mason *et al* (Reproduced with permission from reference [13]. © Springer Science+Business Media New York 1999.)

8.12 Summary

By 2017, over 140 computational models had been reported where rodent models represent 84% of the total amount of computational animal models while 30 computational models were developed for 13 types of animals except rodents, including trout, crab, frogs, flatfish, bees, deer, canines, earthworms, goats, ducks, monkeys, pigs and rabbits. They were adopted in preclinical imaging physics, non-ionizing and ionizing radiation dosimetry, medical physics and environmental radiation protection studies. In all these non-rodent animal models, 43% of the total models are voxel-based models, 33% of are stylized models and 24% of are hybrid models. Computational animal models can be modified to provide a large number of idealized subjects for ionizing or non-ionizing radiation dosimetry, thus reducing the number of subjects needed in animal studies and easing the ethical controversy around animal experiments. However, although the non-human primate is considered to be the gold standard animal model to evaluate the response of the human body to different radiation exposures, only a few computational models of non-human primates have been reported in the literature. Computational models of some laboratory animals, such as the ferret, are still unavailable. Thus, the development of new computational animal models would be welcomed for radiation dosimetry.

Acknowledgements

This work was supported by the Swiss National Science Foundation under grant SNSF 31003A-149957 and the Swiss Cancer Research Foundation under Grant KFS-3855-02-2016.

References

[1] Prise K M, Schettino G, Folkard M and Held K D 2005 New insights on cell death from radiation exposure *Lancet Oncol.* **6** 520–8

[2] Kim J H, Jenrow K A and Brown S L 2014 Mechanisms of radiation-induced normal tissue toxicity and implications for future clinical trials *Radiat. Oncol. J.* **32** 103

[3] Williams J P, Brown S L, Georges G E, Hauer-Jensen M, Hill R P and Huser A K *et al* 2010 Animal models for medical countermeasures to radiation exposure *Radiat. Res.* **173** 557–78

[4] Singh V K and Seed T M 2017 A review of radiation countermeasures focusing on injury-specific medicinals and regulatory approval status: Part I. Radiation sub-syndromes, animal models and FDA-approved countermeasures *Int. J. Radiat. Biol.* **93** 851–69

[5] Woodhead D 2002 Protection of the environment from the effects of ionising radiation *J. Radiol. Prot.* **22** 231

[6] ICRP 2008 Publication 108: Environmental protection – the concept and use of reference animals and plants *Ann. ICRP.* **38** 1–242

[7] Caffrey E and Higley K 2013 Creation of a voxel phantom of the ICRP reference crab *J. Environ. Radioact.* **120** 14–8

[8] Hess C A 2014 Monte Carlo derived absorbed fractions in a voxelized model of a rainbow trout *MS Thesis* Oregon State University

[9] Martinez N, Johnson T, Capello K and Pinder J 2014 Development and comparison of computational models for estimation of absorbed organ radiation dose in rainbow trout (Oncorhynchus mykiss) from uptake of iodine-131 *J. Environ. Radioact.* **138** 50–9

[10] Martinez N, Johnson T and Pinder J 2016 Application of computational models to estimate organ radiation dose in rainbow trout from uptake of molybdenum-99 with comparison to iodine-131 *J. Environ. Radioact.* **151** 468–79

[11] Caffrey E A 2012 Improvements in the dosimetric models of selected benthic organisms *MS Thesis* Oregon State University

[12] Gomez Fernandez M E 2016 Creation and application of voxelized dosimetric models: An evaluation of dose conversion factors in Apis Mellifera *MS Thesis* Oregon State University

[13] Mason P, Ziriax J, Hurt W and D'Andrea J 1999 3-dimensional models for EMF dosimetry *Electricity and Magnetism in Biology and Medicine* (Berlin: Springer) pp 291–4

[14] Toivonen T, Toivo T, Pitkäaho R, Puranen L, Silfverhuth M and Mennander A *et al* 2008 Setup and dosimetry for exposing anaesthetised pigs *in vivo* to 900 MHz GSM mobile phone fields *Bioelectromagnetics* **29** 363–70

IOP Publishing

Computational Anatomical Animal Models
Methodological developments and research applications
Habib Zaidi

Chapter 9

Simulation tools used with preclinical computational models

Pedro Arce, Juan-Ignacio Lagares, Josep Sempau and José-María Gómez-Ros

9.1 Introduction

Although a few studies took place in the 1990s, it was not until the beginning of the 21st century that the use of computational animal models for preclinical studies found a wide range of applications. These applications may be grouped in three fields: medical imaging, ionizing radiation dosimetry, and nonionizing radiation dosimetry. Since the first studies were done, the computational tools used with these models have evolved, especially in the treatment of ionizing radiation. While for nonionizing radiation the tools commonly used are those using the finite-difference time-domain (FDTD) technique, in the field of ionizing radiation, the increase in computer speed has driven the change from the use of relatively simple analytical calculations, like point-source kernels for dose calculations or simple ray tracing for analysis of detector behavior, to an extended use of Monte Carlo simulation. This change has been accompanied by the tendency to use voxelized animal phantoms instead of stylized phantoms based on simple geometrical solids.

9.2 Tools used for simulation

9.2.1 Tools used for ionizing radiation simulation

There are two main applications in the field of ionizing radiation simulation for preclinical studies. The first one relates to the calculations for ionizing radiation dosimetry, where the use of animal models typically serve to calculate organ absorbed dose quantities from internal radioisotopes or external electron or photon beams, while the second field studies the effect of various parameters involved in tomographic imaging.

In ionizing radiation dosimetry, there is a wide range of radionuclides used in mice or rats, including but not limited to C-11, N-13, O-15, F-18, P-32, Cu-64, Ga-68, Sr-89, Y-86, Y-90, Tc99m, In-111, I-123, I-124, I-125, I-131, Pr-143, P-149,

doi:10.1088/2053-2563/aae1b4ch9

Sm-153, Ho-166, Er-169, Lu-177, Re-188, Tl-201, and also beams of electrons or photons of fixed energy, ranging from 0.01 to 10 MeV, under ideal irradiation conditions (left lateral, right lateral, dorsal–ventral, ventral–dorsal, or isotropic irradiation directions). The usual parameters calculated from these sources are organ absorbed fraction (AF), that is the proportion of energy deposited in the target organ released in source organs, and *S*-value, that is the dose rate in the target organ per unit activity. It is generally accepted by the scientific community that the Monte Carlo codes offer the best accuracy for the calculation of deposited energy or dose, although these calculations are relatively slow when compared with other techniques. For these reasons, (and while some studies can still be found that use the point-source kernel method to calculate organ doses) presently, these studies are usually done with a Monte Carlo simulation. It is important to note that the input parameters used for the description of small animals, such as model resolution, organ segmentation, tissue density, and spatial sampling, have proved to have a substantial influence on absorbed dose calculations when using the Monte Carlo simulation [1].

In medical imaging, the computational models serve to provide precise information to evaluate the impact of physical degrading factors inherent to the imaging process, to assess different design concepts and performance of medical imaging systems, or to advance the development and validation of new image segmentation, registration, reconstruction and processing techniques. The variety of imaging systems studied is wide: positron emission tomography (PET), single photon emission computed tomography (SPECT), bioluminescence tomography (BLT), and fluorescence molecular tomography (FMT). Monte Carlo codes are also the most commonly used calculation tools for simulations in this field. These codes can precisely simulate all the physical aspects of the animal phantoms and the detectors, and they are usually combined with reconstruction software that reproduces, in detail, the machine reconstruction of images from the detector signal. Nevertheless, in the past decade, it is also possible to find some studies where a simple ray tracing technique was used to model the point spread function response of SPECT detectors.

The point-kernel method is a macroscopic approach where the effects of radiation interaction in matter are described using macroscopic linear attenuation factors and adding build-up factors to account for scattered radiation. As consistent scattered radiation accounting cannot be achieved within a macroscopic approach, a common practice is to use semi-empirical relations such as the Berger formula, Taylor formula, etc. The radiation source volume is cut up into elementary cells (point kernels) and for each point kernel, the contribution to the dose rate at the detection point for each radiation energy is calculated. Point-source kernels are assumed to be independent so that the total dose rate at the detection point is obtained by integrating the contribution of each source volume summing over the energies of the radiation spectrum.

As Monte Carlo is currently the generally accepted tool to model ionizing radiation studies with animal phantoms, we will describe this technique in detail and analyse several codes that are offered to the user, comparing the capabilities and drawbacks as well as the performance of the major Monte Carlo packages.

9.2.2 Tools used for nonionizing radiation simulation

Many animal models have been used in experiments involving the assessment of the biological effects of radio frequency (RF) based devices, such as electromagnetic waves, wireless local area networks, high-field MRI, and, more recently, personal cellular phones. Some studies calculate the heat increase due to the RF exposure or high-field MRI either by using the bioheat equation [2] on a numerical phantom or developing an FDTD-based thermal model to evaluate the temperature elevation due to electromagnetic energy deposition in high-field MRI [3]. But, practically all studies in this field are interested in the calculation of the localized specific-absorption-rate (SAR), that is, the rate at which energy is absorbed by a body when exposed to an RF electromagnetic field and its comparison with empirical measurements in animal experiments.

Except for a few cases where the SAR is calculated with a simplified method, SAR is calculated by using the FDTD technique. FDTD is a numerical analysis technique used for modeling computational electrodynamics (finding approximate solutions to the associated system of differential equations). The time-dependent Maxwell's equations in partial differential form are discretized using central-difference approximations to space and time partial derivatives. The resulting finite-difference equations are solved in either software or hardware in a leapfrog manner: the electric field vector components in a volume of space are solved at a given instant in time, then the magnetic field vector components in the same spatial volume are solved at the next instant in time, and the process is repeated over and over again until the desired transient or steady-state electromagnetic field behavior is fully evolved.

As the complex structures of biological beings greatly influence the SAR distribution, it is necessary to make a detailed model of the animal under study. FDTD is usually done in a structured Cartesian grid, so it seems natural to make a 3D grid similar to the voxelized phantoms used for the absorbed dose calculation. However, resolutions under a millimeter are difficult to achieve with the FDTD method, so the phantoms used for this application are no more precise than 1–2 mm. On the other hand, the precision needed for these studies is not as stringent as that required for ionizing radiation dosimetry studies in small organs and, therefore, this resolution is quite satisfactory.

9.3 The Monte Carlo simulation method

The Monte Carlo method is a numerical technique used to solve mathematical problems by using random numbers (its name is taken from Monaco's capital city, well known for gambling). In mathematics and physics, there are many problems that can be expressed in the form of differential or integral equations. Often these equations cannot be resolved by applying analytical methods without introducing approximations that reduce the accuracy of the results. For these problems, the solution may be to use numerical methods, like the Monte Carlo method. This is often the case in problems related to the transport of ionizing radiation and its interactions with matter, which are well described by statistical equations, so that they can be described by probability distribution functions. To solve these problems, particles follow a step-by-step process simulating the real trajectory, and the probability of interaction with matter through one of the possible processes (e.g. Compton scattering, Rayleigh scattering, photoelectric effect or e+e− pair production for a gamma, bremsstrahlung, or ionization for an electron) is sampled by generating a random number according to the corresponding distributions. When one interaction happens, other random numbers are generated to sample the energy loss, the dispersion angle, or the energy and momentum of the secondary particles created in the interaction.

The interaction of charged particles usually happens at very short distances and creates many small secondary electrons, and photons, as well as many atoms are left in excited states. The simulation of this large number of interactions would require a prohibitive amount of time, but indeed this is not necessary unless one is interested in microdosimetry studies. Therefore, all the Monte Carlo codes simulate the charged particle interactions through different variations of the condensed history technique [4], in which the path of the electron is broken into a series of steps and the effects of a large number of individual interactions occurring during the step are grouped together. One grouping accounts for the large number of deflections caused by elastic scattering by using multiple-scattering theories such as that of Molière [5–7] or by Goudsmit and Saunderson [8]. The other major grouping accounts for the large number of small energy losses. Most of the major Monte Carlo codes today are based on what Berger calls class II models, in which individual interactions change the energy and direction of the primary charged particle only when they create ionization knock-on electrons or bremsstrahlung photons above certain energy thresholds, and the effects of secondary particle production below these thresholds are grouped together. All energy that the primary particle loses between interactions without creating a secondary particle is summed together in what is called 'deposited energy', which constitutes the quantity from which the dose is calculated (dividing it by the mass of the volume where the interaction happens). Monte Carlo N-Particle (MCNP) is an exception to this, as it uses a Class I condensed history implementation, where electron transport is performed on a predetermined step-size grid. This allows the calculation of the multiple elastic scattering distribution, for arbitrary complicated cross sections, for the step-size grid used with a reasonable amount of pre-calculated data.

One important aspect in Monte Carlo simulation is the determination of the precision associated with the calculation. As the calculations of macroscopic quantities are obtained by averaging the values of a large number of particles, the precision will be given by the number of trials or simulations used. This precision will be, in general, proportional to the square root of the number of trials, or, in other words, to diminish its value by an order of magnitude it is necessary to increment the number of simulations by two orders of magnitude. As this often implies a high computational cost, even in the most advanced computers, it is crucial that the user decides which uncertainty is desired before starting a simulation. This can be done by simply running the setup with a small number of trials and looking at the magnitude of the uncertainty. It should, however, be considered that the rule of diminishing the uncertainty with the square of the number of trials is only a general rule, only valid asymptotically, that is, when the number of trials tends towards infinite, and therefore it may not be suitable for the user's application.

In the case of the Monte Carlo simulation of ionizing radiation, the probability functions that describe each process are taken either from a set of experimental measurements or from analytical expressions describing them. Therefore, the precision is also limited by the uncertainty in the measurements or in the approximations often used to derive the analytical equations. In general, a precision less than 1% cannot be expected.

9.3.1 Monte Carlo simulation of computational phantoms

The simplest computational phantoms are the so-called 'stylized' or 'mathematical' phantoms. In these phantoms, the different organs are defined as simple solids like cubes, prisms, spheres, cones, ellipsoids or as Boolean operations of them (i.e. unions, subtractions, or intersections). Even with complicated and carefully designed Boolean operations, these kinds of phantoms are not anatomically realistic. On the other hand, they have the advantage that any Monte Carlo code can simply model these types of phantoms, usually through the technique called constructive solid geometry (CSG), which models the solid surfaces by quadric equations, making the simulations in these geometries very efficient.

A more precise way to simulate an animal body is by using 'voxelized' or 'tomographic' phantoms. The body geometry is divided into a set of contiguous cuboids, all of the same dimension, and each organ is assigned to a set of voxels. There is also the possibility of assigning to each voxel not only a different organ, and therefore material, but also a different density, by using a table to convert the Hounsfield number (which corresponds to the linear attenuation coefficient) of a CT scan to a material and material density [9, 10]. Nevertheless, if a thorough identification of each organ is done, it is not clear that assigning different densities to the material of an organ has a relevant advantage in the case of animal phantom studies, except maybe for the lung. The regular geometry of a voxelized phantom is very simple for Monte Carlo codes to handle, although a large number of voxels may pose a problem of memory or CPU time (see section on code comparison below). This precludes the use of a small voxel size, which implies that some organs may not be described with enough fidelity, especially for thin or small organs such as the skin, eye lens, ribs or bone marrow. Furthermore, a voxel phantom is a rigid structure, and it is very difficult to change an organ size or shape to simulate the anatomical variability associated with the desire to simulate different individuals or changes due to bodily functions like respiration or heartbeats.

The most powerful techniques to simulate animal geometry are the BREP (boundary representation) methods, configuring what are commonly referred to as 'NURBS', 'mesh,' or 'BREP' phantoms. In this technique, the exterior of an object is defined as NURBS (non-uniform rational b-spline), which gives very smooth surfaces. The object surfaces can alternatively be represented as polygons whose vertices are defined by a set of coordinate values x, y, and z. Unlike the other two representations, BREP is very flexible and several software tools, usually commercial, allow one to deform the objects in many ways: extrusion, chamfering, blending, drafting, shelling, and tweaking. Despite its advantages, there is no Monte Carlo simulation code that can manage BREP solids represented as NURBS. If the BREP solids are all represented by triangular or quadrangular facets, some codes have the functionality to simulate it, and it has been demonstrated that the use of a faceted phantom with Geant4 makes the simulation much faster than the use of voxelized phantoms [11].

As mentioned above, the most common magnitude of interest in the preclinical use of computational animal phantoms is the absorbed dose to an organ, usually the cross-organ (dose to an organ from particles originating in another organ) or self-organ (dose to an organ from particles originating in the same organ). Therefore, it is important that a Monte Carlo code provides the functionality of handling organs, either to define particle sources or to restrict the dose calculation to a given organ.

9.3.2 Monte Carlo simulation of medical imaging detectors

The simulation of a medical imaging detector is a complex task. The geometry of a detector can be described with enough accuracy using simple solids, but the complexity lies in simulating the detector effects. If the detector is made of scintillation crystals, as the majority of tomographic imaging detectors today are, the signals are created when a charged particle produces fluorescence photons in the crystals, which create an avalanche of electrons in the photomultiplier that is transformed into electronic signals, which are then reconstructed by the detector software to convert them to energy, position, and time. It is not necessary to simulate this process in detail, but a Monte Carlo code can summarize it by calculating the deposited energy into signals and summarizing these effects as energy, position, and time resolution. This is, however, insufficient, as other effects must be simulated, like the grouping of crystals as well as the pile-up and the effect of the detector electronics' dead time. Lastly, the grouping of signals and classification as 'good' events, that is, those that pass to the reconstruction code to create the image requires a non-negligible amount of coding. All this would be enough to reproduce the behavior of the detector, but the simulation code should add some extra utilities to help better understand the physics of the process, like the creation of scatter or attenuation maps or algorithms to identify the first interaction when a particle has several interactions in the detector crystals (as the first interaction is the one that better serves to extrapolate the position of the particle source).

9.4 Monte Carlo packages for preclinical studies

There are several Monte Carlo packages publicly available that can be used for treating ionizing radiation. Among them, the major codes by their number of scientific citations are EGS4 [12]/EGSnrc [13]/EGS5 [14], FLUKA [15, 16], Geant4 [17–19], MCNP5 [20]/MCNPX [21]/MCNP6 [22], and PENELOPE [23–25]. Any of these can read voxelized phantoms and calculate a map of absorbed doses. Nevertheless, none of the major Monte Carlo packages can manage organs delineated in a DICOM image without a major intervention by the user in the form of writing many lines of code; this also implies a deep understanding of the Monte Carlo code itself. Only Geant4 can simulate directly an isotope with all its decay chain, while for the other codes it is necessary to simulate a single particle with a certain initial energy distribution that can be obtained from an external source or, in the case of PENELOPE, from a package provided with the code. It is also quite complicated to simulate a medical imaging detector, having to code all the physical effects associated with the signal detection: like energy, position, or time resolution, association of energy depositions into hits, merging of signals from different particles when they are not distinguishable by the hardware, the effect of the dead time of the detector units, etc.

All these functionalities are usually provided by some applications that have been created to facilitate the use of these codes. Concerning the dose calculations in organs, we can cite PRIMO [26] (based on PENELOPE), GAMOS [27], GATE [28], or TOPAS [29] (based on Geant4). Concerning the simulation of medical imaging, we can cite PENELOPET [30] (based on PENELOPE) and GAMOS or GATE (based on Geant4).

9.4.1 EGS

EGS (electron gamma shower) is known as a general-purpose package for the Monte Carlo simulation of the coupled transport of electron-positrons and gammas developed at the Stanford Linear Accelerator Center. The history of EGS starts at the beginning of the 1960s with the codes written by Zerby and Moran [31–33] at the Oak Ridge National Laboratory of the USA and the codes written in FORTRAN by Nagel [34–36], one of which is named SHOWER1. The first EGS code (EGS1) was written in the 1970s by Nelson and Ford and a few years later EGS2 was written in MORTRAN. MORTRAN is an extended version of FORTAN that has to be transformed to a standard FORTRAN before compilation. The MORTRAN code was used again in versions EGS3 and EGS4. Today, the latest version of the package, EGS5, is written in FORTRAN.

In 2000, the first version of the code EGSnrc was published, based on EGS4, by I Kawrakow *et al* [37] at the National Research Council of Canada. The objective was to enhance the capabilities of EGS4 specifically for low energies. This package is written in MORTRAN3.

Both codes, EGS (in its new version) and EGSnrc, have increased capabilities but an important difference marks both codes: EGSnrc has continuous releases and was created to help solve medical physics problems and, in fact, evolved in parallel with another Monte Carlo package, BEAM, that is now distributed and integrated in the same package as EGSnrc (BEAM will be introduced below). In fact, both EGS5 and EGSnrc have been used to solve problems in the medical physics field and in preclinical and clinical problems and applications, but, given the advantages, features, and the number of works published with EGSnrc, we will take the EGSnrc package as a reference.

Codes and geometries

EGS codes proved the base for transport particles, but the geometry, sources and scores have been developed for the final users in a separate code.

One of the advantages of EGSnrc is that it is distributed with a variety of user codes which are able to solve a large number of problems in the medical physics field. All user codes work in a similar way: after being compiled, they are used by writing an input file defining the particle source, geometry and, transport parameters, and a pegs4 file. The pegs4 file is a pre-processed file with material data used in the EGSnrc codes and can be generated with the tool provided in the EGSnrc distribution. In 2004, EGSnrcMP (EGSnrc multi-platform) was released; this version can be run in UNIX and Windows systems and includes all EGSnrc user codes and a graphical user interface where all parameters may be filled in or selected.

Each of the basic codes distributed with EGSnrc is meant to solve a specific problem in a specific geometry. DOSRZnrc scores deposited doses in cylindrical geometry, FLURZnrc scores particle fluence, CAVRZnrc scores some quantities used in dosimetry for ion chambers, SPRRZnrc calculates the Spencer–Attix spectrum averaged stopping-power ratios for arbitrary media, CAVSPHnrc, that it is the equivalent to CAVSRZnrc and EDKnrc, calculates the spherical geometries energy deposition kernels for photons or electrons, which are forced to interact at the center of a spherical phantom or in a specific region.

A separate mention is required for the code DOSXYZnrc. It is a general-purpose code to calculate absorbed dose in a 3D Cartesian geometry, i.e. a voxel phantom. The phantom can even have a different thickness for each *XY*, *XZ* or *YZ* plane, and a unique material may have different densities in different voxels. The package includes an application (ctcreate) to export a CT with conversion data to a phantom format to be read for DOSXYZnrc. It also includes the necessary elements to work with parallel computing system and options to analyse and add results.

BEAM, known today as BEAMnrc, is a Monte Carlo simulation system for modeling radiotherapy sources. It was developed as part of the OMEGA project. The main aim of the OMEGA project was to develop a full Monte Carlo radiotherapy treatment planning system. It was started at the University of Wisconsin in the middle of the 1990s. BEAMnrc was developed under the EGSnrc system and gives the final user an easy interface to generate the output phase-space of a radiotherapy treatment using linacs ^{60}Co units or x-ray tubes. For these purposes, the package offers a limited, but in most cases sufficient, list of sources and geometry modules to simulate the most common commercial linacs including multi-leave collimators; it is written in MORTRAN3. The user has the option to use a graphical interface to select the necessary geometrical module and compile a specific executable file for those geometry modules. After that, under the graphical interface, the users are able to introduce the parameters to define the specific data for those geometries, select a specific source, the plane or planes to obtain a phase-space, and the transport parameters, including variable reduction techniques such as bremsstrahlung splitting, electron range rejection, photon forcing, and others.

Due to the difficulty of writing new user codes and the limited geometries, sources, and scores available in each distributed code, a new C++ geometry class was added to the EGSnrc package. This new library provides an easy way to model a wide range of geometrical structures in combination with a wide set of sources and a set of basic scoring classes like dose deposited. This gives new possibilities like simulating a PET geometry together with a voxel phantom that is not available in the basic user code provided. The final user has to complete a basic C++ program, and the source and geometry have to be written in a text input file. The text input file has a simple and logical structure, and many operations with the basic geometry can be done, like adding or including geometries one inside another. The geometry classes available are planes, concentric cylinders, concentric spheres, parallel cones, simple cones, cone stacks, planes, boxes, prisms, pyramids, and also more complex ones: NDgeometry, XYZGeometry, DeformedXYZ, EnvelopeGeometry, TransformedGeometry, UnionGeometry, StackGeomtry, CDGeomtry (combinational dimension), and XYZRepeater (see EGS user's manual for a detailed description of these classes).

Particle generator
All user codes provide a variety of primary particle geometries like punctual sources, parallel cylinder beams with angle on-axis or off-axis, or parallel rectangular beams with angle on-axis or off-axis. The phase-space utility includes the option to use the IAEA phases' spaces format and change the angle of the particles read. It is possible to select a mono-energetic source or an energy spectrum given by the user, and it is also possible to select one of the spectra provided together with the code, which includes the decays of several isotopes.

Physics
EGSnrc provides physics for electron–positron and gamma transport from 1 keV to several hundred GeV, and multiple options to select various models for different interactions, algorithms, or cross sections for particle transport.

The user must choose between different options depending on the problem, as some options are more accurate than others. Several options can be activated or not, Rayleigh scattering, atomic relaxations, electron impact ionization, or the appropriate method to evaluate the 'pair angular sampling' can be selected. One interesting option is that the photon cross sections library can be selected between XCOM, Evaluated Photon Data Library (EPDL), Storm-Israel, or PEGS4.

But maybe, one of the most characteristic physics codes that distinguishes EGS is the electron step algorithm named PRESTA I (parameter reduced electron step transport algorithm) and its newer version, PRESTA II, both with the options to choose electron transport options and the incorporation inside of the boundary crossing algorithm, eligible as a PRESTA I or EXACT.

Complementary tools
EGSnrc provides complementary tools with the user codes and some of them are mentioned above as ctcreate. Other important tools included in the OMEGA project are dosxyz_show that visualizes dose distributions obtained with DOSXYZnrc over a voxel phantom, Statdose that visualizes rebin dose distribution obtained with

DOSXYZnrc, BEAMdp that analyses phase spaces, and addphsp to add phase-space files.

9.4.2 Geant4

Geant4 (geometry and tracking, version 4) is a large-scale object-oriented toolkit written in C++ by over one hundred collaborators that are geographically distributed. Although initially created for high-energy particle physics, it was extended early on to the lower energy region needed for space and medical physics. Its current applications extend to many physics fields including particle physics, nuclear physics, accelerator design, space engineering, high-energy astrophysics, medical physics, etc, making it the most-cited Monte Carlo code.

Geant4 is considered by its authors to be a toolkit, as it provides a diverse, wide-ranging yet cohesive set of software components that can be employed in a variety of settings. It offers a wide functionality for the simulation of the geometry, including CSG solids, voxelized phantoms, as well as faceted geometries. Practically all kinds of particles can be simulated, not only gammas, electrons, and positrons, but also protons, neutrons, ions, and all kinds of mesons and baryons and their antiparticles. Concerning the physics, there is a wide range of processes available, ranging from a few eV for microdosimetry studies to the PeV realm of cosmic muons, and offering, in many cases, several competing physics models.

One of the caveats of GEANT4 is the need to write the user code in C++. To avoid a long and steep learning curve, several applications have been created based on GEANT4 to facilitate the use of GEANT4. These applications provide a simple scripting language, that means the user does not need to know C++ and the inner details of the GEANT4 code itself, And these applications also provide a set of utilities that spare the user the task of writing many hundreds of lines of code to obtain the desired results. Three of these applications stand out in the field of medical physics: GAMOS, GATE, and TOPAS. Related to the specific field of simulations for preclinical studies with animal phantoms, the direct use of Geant4 without one of these applications becomes a huge task, as the user would have to write many hundreds of lines of C++ code. This is the reason why, in the following, we describe the utilities that Geant4 offers together with those offered by one of these applications, GAMOS, when the direct use of Geant4 would be a big task.

Physics

Geant4 offers a modular approach to simulate the physics of the particle interactions with matter. There is no default physics list, but the user must define which particles will be included in the simulation, which processes are defined for each particle (e.g. ionization, bremsstrahlung and multiple-scattering for electrons, or Compton interaction, Rayleigh interaction, photoelectric effect and creation of e+e− pairs for photons), and for each process the model must be selected among the several that Geant4 usually offers.

For electromagnetic interactions, there are three sets of physics models, one named Standard and two others designed for low-energy: Livermore and Penelope. The Standard package [38] provides a variety of models based on an analytical approach; it is optimized for high-energy physics applications, but it describes the interactions of electrons, positrons, photons, and charged hadrons and ions from 10 PeV down to 1 keV. The Livermore package is based on a parameterized approach exploiting evaluated data libraries (EPDL97 [39], EEDL [40], and EADL [41]) to calculate cross sections and to sample the final state. The so-called Penelope package combines numerical databases and analytical models for the different interaction mechanisms; these models were originally developed for the PENELOPE Monte Carlo FORTRAN code version 2008, but they have been re-engineered into Geant4 with an object-oriented design. A good comparison of the three electromagnetic sets of models can be found at [42].

If the user is interested in microdosimetry, Geant4 provides a sub-library of very low-energy models; the Geant4 DNA project that simulates radiation effects involving physics and chemistry at the sub-cellular level, down to 7 eV in water. Also, many different models for hadronic interactions of all kinds of particles are available.

For more details on the physics processes and models available in Geant4, refer to the Geant4 Physics Reference Manual [43].

Geometry

The geometry module can simulate from very simple solid shapes like boxes, trapezoids, spherical and cylindrical sections, to more complex ones like torus, ellipsoid, tetrahedron, extruded solid, twisted trapezoid, with each having their properties coded separately, in accord with the concept of CSG. Another way of obtaining solids is by Boolean combination (union, intersection, and subtraction), where one of the components may have an optional transformation relative to the other. Also, several applications outside Geant4 are available to convert geometries from computer-aided design (CAD) systems, via the ISO STEP standard, to triangular or quadrangular facets that can be simulated by Geant4. The placement of repetitive structures can be represented by specialized volumes, replicas and parameterized placements, with sometimes an enormous saving of memory. To simulate a voxelized phantom, Geant4 proposes an optimized parameterization that includes a fast navigation in these types of regular geometries [44], offering also the possibility to skip voxel frontiers when both voxels are made of the same material. Although a detector is naturally and best described by a hierarchy of volumes, efficiency is not critically dependent on this because an optimization technique, called smart voxelization, allows efficient navigation. An inconvenience of Geant4 is that it is not able to simulate a material with different densities but it must create one different material for each material–density combination. Together with each material, Geant4 must store, in memory, the physics tables for each process, which could use a large amount of memory if many different densities are present in the phantom. To avoid this memory problem, Geant4 proposes a phantom reader that merges materials whose density is inside an interval defined by the user.

Particle generator

Geant4 provides a general class to define the generator of primary particles. The available particles include 2831 isotopes, with their decay tables, based on the

ENDSF data library. It is also possible to simulate isotopes in an excited state as primary particles that will automatically decay into the ground state and emit a gamma with the excited isotope energy. The difficulty lies in the fact that Geant4 does not provide an example of how to create particle distributions suited to simulate an activity distribution coming from a PET scanner or to select the origin of the particles in a given organ delineated in the phantom. GAMOS does provide these distributions and it is able to read PET DICOM and RTSTRUCT DICOM files and use them as particle generator distributions, all by just adding a few lines to the input script. How to do this is explained in the GAMOS DICOM tutorial that is provided with the GAMOS code. Moreover, GAMOS has several dozen distributions of position, direction, energy, and time available.

Regarding the simulation of nuclear medicine detectors, Geant4 has the advantage with respect to other codes because it is able to simulate any time distribution and several particles in the same event (we understand 'event' as the initial particle track plus all secondary tracks). This, together with the possibility to simulate radioactive ions with their decay chain, allows for a detailed simulation of the pile-up as well as the effect of dead time on the detector electronics. Although Geant4 does not provide an example to simulate a nuclear medicine detector, GAMOS provides the full functionality needed to simulate PET, SPECT, or Compton Camera detectors.

Scoring

Geant4 offers the possibility to assign one of several scorers to one of several volumes. Among the scorers, a user can select the energy deposit or dose deposit scorers, which usually are enough for studies in animal phantoms for preclinical use. The caveat of the Geant4 scoring utility is that it does not manage RT structures, but the user must develop their own code to do the scoring on a given organ. This functionality is available in GAMOS, which, in addition, offers the user the possibility of doing all the desired scoring in a few lines of the user script.

9.4.3 MCNP

MCNP (Monte Carlo N-Particle) is a general-purpose Monte Carlo code that has been developed at Los Alamos National Laboratory (USA) for over 50 years. The latest version of the code is MCNP6, that merged former MCNP5 and MCNPX codes, and the three that are distributed by RSICC. Version MCNP5 was released in 2010 and can be used for neutron, photon, electron, or coupled neutron/photon/electron transport. MCNPX 2.7 extended the capability the of continuous-energy transport of 34 particles and light ions using cross-section libraries when available and models otherwise. In both codes, MCNP5 and MCNPX, the photon and electron cut-off energy is 1 keV. MCNP6 extends the energy range of electrons down to 10 eV and photons down to 1 eV, also allowing the simulation of electrons event by event [22, 45].

Geometry

Geometry modeling in MCNP is done in terms of volumes (cells) bounded by surfaces and logical operations (intersection, union, complement) of these volumes. Possible surfaces include first- and second-order-degree planes, cylinders, spheres, cones, quadric surfaces (ellipsoids, paraboloids, hyperboloids), and tori. Up to 10^8 surfaces and cells can be defined in MCNP6 depending on the available memory [45]. Moreover, the repeated structure feature makes it possible to describe only once the cells and surfaces of any structure that appears more than once in a geometry.

Voxelized geometries (voxel phantoms) can be implemented in MCNP by using 'UNIVERSE', 'FILL', and 'LATTICE' cards to define the voxel phantom geometry as a rectangular parallelepiped filled by a lattice of $n_x \times n_y \times n_z$ cubic elements (voxels), with every one identified by a number (universe). The sequence to fill the lattice is given and then every universe is filled with the appropriate material with a given density. Thus, different identification numbers are used for the Universes (each one corresponding to a given organ) and for the materials. Therefore, the same material can be used to fill different organs either with the same or different density (for instance, a simplified model for photon/electron transport can be built from a tomographic image just considering liquid water or soft tissue as a single material for all the voxels, and a set of discrete density values calibrated according to the Hounsfield numbers).

Particle generator

The source term in MCNP is defined by specifying the parameters of the emitted particles using either fixed values or probability distributions in the 'SDEF' card. MCNP provides different standard tallies to score fluence (using track length estimators and partially deterministic methods), absorbed dose (in the case of photons and neutrons, both with kerma approximation or considering the transport of secondary charged particles), and related quantities. In addition, it is possible to create user-defined subroutines, 'SOURCE' and 'TALLYX'. For voxel phantom geometries, the source can be distributed by voxels or uniformly in the voxels corresponding to a given organ.

Scoring

Scored quantities (e.g. deposited energy or absorbed dose) can be tallied either in the voxels or in organs. In this case, both +f6 and *f8 tallies have been employed. Tally +f6 is a track length heating estimator that accounts for energy deposition from all particles in the problem to calculate the absorbed dose. The energy of secondary particles is subtracted from the heating values and is handled in the regular process of tracking those particles. Tally *f8 is an energy deposition estimator, only valid for photons and electrons, that directly accumulates the energy deposited in a given volume. In addition, mesh tallies can provide a 3D map of deposited energy using a spatial resolution independent of the voxel lattice.

9.4.4 PENELOPE

PENELOPE is a general-purpose Monte Carlo code system for the simulation of coupled electron–photon transport in the energy range between 50 eV and 1 GeV. It has been developed at the University of Barcelona for over 20 years [23, 24]. Its current version is 2014 [25].

The core of the system is a FORTRAN subroutine library that generates particle trajectories in arbitrary materials. The library is invoked from a steering main program that generates primary particle states (i.e. it defines the source term), controls the evolution of the tracks, keeps score of the quantities of interest (energy deposition, fluence, etc) and, if required, applies variance reduction techniques to speed up the calculation of some ill-conditioned problems. The standard package comes with two main program examples. Other authors have developed alternative main program, such as penEasy [46]. The combined system formed by the standard library and this main program will hereafter be referred to as PENELOPE/penEasy.

Both PENELOPE and penEasy are free and open source codes written in FORTRAN that can be run on any operating system for which there is a FORTRAN compiler, which includes all common UNIX flavors, MS Windows, and Apple's OSX. In particular, the penEasy package includes an executable file to run the code under Windows.

The PENELOPE package is available from the Nuclear Energy Agency (https://www.oecd-nea.org) and, in North America, from the Radiation Shielding Information Center at Oak Ridge National Laboratory (https://rsicc.ornl.gov). The penEasy package can be downloaded from http://www.upc.es/inte/downloads.

Physics

PENELOPE generates a particle track as a sequence of free flights. At the end of each flight segment the particle suffers an interaction with the medium where it loses energy, changes its direction of flight, and, in certain cases, produces secondary particles.

The photon interaction mechanisms considered here are:

(i) photoelectric effect, with total cross sections and partial cross sections for the K, L, M, and N atomic shells that are practically coincidental with those from the EPDL, but obtained with more accurate numerical algorithms and some additional corrections,

(ii) coherent or Rayleigh scattering, with total cross-section obtained from the EPDL,

(iii) incoherent or Compton scattering with a differential cross-section (DCS) obtained from the relativistic impulse approximation that accounts for electron binding effects and Doppler broadening,

(iv) electron–positron pair production with total cross sections obtained by interpolation in a table generated with the XCOM program [47]. The transport of polarized photons is also considered. This also affects the Rayleigh and Compton scattering DCS models.

For electrons (and positrons), the interaction mechanisms are

(i) elastic collisions with a DCS database [48] obtained from Dirac partial-wave cross sections calculated with the program ELSEPA [49],

(ii) inelastic collisions with a DCS obtained from the Born approximation with a generalized oscillator strength model that accounts for the density effect correction. Inner (K, L, and M) shell ionization by electron impact is described independently from (and more accurately than) the inelastic model by combining calculations based on the plane- and distorted-wave Born approximations, depending on the energy of the projectile.

(iii) bremsstrahlung emission described by using a modified Bethe–Heitler model that includes Coulomb and low-energy corrections

(iv) positron annihilation based on the Heitler DCS.

PENELOPE takes into account the emission of characteristic x-rays and Auger electrons that result from vacancies produced in the K, L, M, or N atomic shells.

Photon simulation is performed by using the conventional detailed simulation scheme, that is, all interactions in a photon history are sampled one-by-one in chronological succession. For electron transport, a mixed simulation scheme is used in which 'hard' individual interactions with energy losses or angular deflections larger than certain thresholds are simulated in a detailed way. The global effect of the remaining interactions, called soft, that occur between each pair of successive hard interactions is simulated as a single artificial event, where the particle loses energy and changes its direction of flight according to suitable multiple-scattering theories. The thresholds are determined by four (for each material) user-defined transport parameters. Detailed simulation can be recovered by setting these parameters to zero, although this possibility is only practical at relatively low energies due to a large number of individual interactions that a charged particle undergoes before coming to rest.

Geometry

The PENELOPE package includes subroutines that allow tracking in homogeneous bodies limited by quadric surfaces. Bodies can be grouped into modules that, in turn, can form part of larger modules. The geometry is introduced by the user in a plain text file that can be visualized in 2D and 3D with the help of an auxiliary code included in the package.

PENELOPE/penEasy extends this capability to simulate voxelized geometries, that is, objects described by a large set of homogeneous cuboids or voxels. Each voxel can be assigned an arbitrary mass density, which may be different from the nominal density of the material of which that voxel is made. Additionally, the code permits the combination of quadrics and voxelized geometries, in which a portion of the 'universe' is described with voxels, whereas quadric surfaces are used for the rest. Voxels and quadrics can overlap, a feature that can be used to define complex mixed cases, for example, a small brachytherapy source (limited by quadric surfaces) embedded inside the patient's body (voxelized geometry).

Particle source

PENELOPE/penEasy includes two source models. The first one is used to describe point and distributed sources that emit particles isotropically but restricted to a spherical window defined by an interval in the polar and azimuthal angles, for example, a cone or the whole sphere. The energy spectrum can be defined explicitly by the user or by means of the package PENNUC [50], which simulates random decays of radioactive nuclides using information from the NUCLEIDE database of the Laboratoire National Henri Becquerel. PENNUC is seamlessly embedded in PENELOPE/penEasy.

The second model is a phase-space file (PSF), that is, PENELOPE/penEasy can read an external file containing the initial state of each particle to be simulated. The code can read PSFs in text format or in the format defined by the International Atomic Energy Agency (see https://www-nds.iaea.org/phsp for details).

Tallies and dose scoring

The code includes eleven different tallies such as fluence spectrum, energy deposition, dose distribution in voxelized geometries, and pulse height spectrum. Tallies can be cloned easily to, for example, tally pulse height spectra in two different detectors simultaneously. Also, the code is structured in a way that facilitates the development of new tallies (and source models) by advanced users.

9.5 Comparison of performance of Monte Carlo packages

We compare here the performance of four of the major Monte Carlo codes, EGSnrc, Geant4, MCNP6, and PENELOPE, in simulating animal phantoms, focusing on two aspects: the amount of memory and CPU that they need to simulate a large number of voxels, and the comparison of the physics performance using a dose map calculated in all the phantom voxels. Given the difficulty in directly using the codes, we will use the penEasy application instead of PENELOPE, GAMOS instead of Geant4, and EGSnrc plus DOSXYZnrc instead of EGS. For MCNP, we use the latest version MCNP6, as previous versions cannot simulate a phantom with such a big number of voxels unless the user modifies the code.

For each code, we have selected the default options, except where we think the performance may improve without spoiling the results. The versions of the code and physics options used in this comparison are:

- EGSnrc version 4 and DOSXYZnr version 1.54, with the cross sections calculated by PEGS. We have made two changes in the code before recompiling it: we modified the 'dosxyznrc_user_macros.mortran' file so that the dose for voxels with an error smaller than 50% of the value are also written, and the file 'dosxyznrc.mortran', so that it can simulate more than nine materials.
- GAMOS version 5.2.0 based on geant4.10.03.p01 using the standard electromagnetic package, as it is considerably faster than any of the other two options, Livermore and PENELOPE. Also, for Geant4, we have set the option to skip voxel frontiers when they have the same material (see [44]) as this saves a factor of more than three in CPU time.
- MCNP6 version 1, using the cross-section libraries mcplib84 for photons and el03 for electrons [51].

- PenEasy version 2015-05-30 based on PENELOPE version 2014.

The mandatory production threshold is set to 10 keV for gammas and electrons. For the cut to kill all particles whose energy is below a value, we have also chosen 10 keV. This second cut is mandatory in all codes except for the case of Geant4; setting this code to such a low value does not imply any significant improvement in performance.

The animal phantom chosen for the comparison is MOBY [52], which is the most-cited animal phantom in scientific publications. MOBY is a realistic and flexible 4D digital mouse phantom built from high-resolution 3D magnetic resonance microscopy data (figure 9.1). The organ shapes are modeled with NURBS surfaces, but, as the Monte Carlo codes cannot simulate NURBS solids, we used the utilities provided with the phantom to convert it to a voxelized one, using the attenuation map to convert it to material and material density with a simple conversion table.

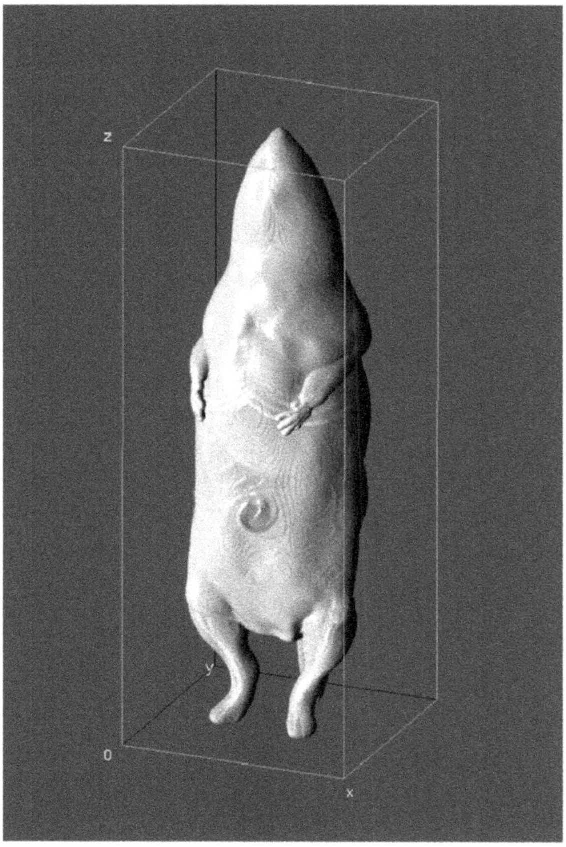

Figure 9.1. 3D view of the simulated MOBY voxel phantom displayed using ImageJ software.

9.5.1 Memory consumption

Taking the default MOBY voxel size of $0.145 \times 0.145 \times 0.145$ mm^3 we obtain a total of $256 \times 256 \times 744 = 48\ 758\ 584$ voxels. This large number of voxels is a challenge for these Monte Carlo codes, but it is indeed not uncommon for animal phantoms to have such a large number of voxel dimensions. To understand the limits of the Monte Carlo codes, we have first made a study on the amount of memory needed to store the geometry in memory, as well as to store the quantities needed to calculate the absorbed dose (not only the dose, but also the dose squared, to take into account the correlations between the doses deposited by different tracks or events, plus some temporal dose variables). We have made all benchmarks using the same machine, running a Scientific Linux distribution of 64 bits.

The comparison of the four codes can be seen in table 9.1. In the case of EGS, we do not quote a number for the phantom geometry alone because the code automatically builds a dose matrix together with the voxel matrix. For GAMOS/Geant4, it is worth mentioning that the dose matrix is filled dynamically, using a standard template library (STL) map, so that the amount of memory increases proportionally to the number of voxels where a particle has deposited some energy; this saves some memory if the fraction of filled voxels is small, at the cost of increasing it if this fraction is big. GAMOS proposes an alternative solution, using STL vectors, which occupy only 8 bytes compared to the 48 that occupy an STL map, but it needs to reserve the memory for the full dose matrix at the beginning. In the case of the MOBY phantom, this option does not represent a big advantage, as about two-thirds of the voxels belong to the space surrounding the mouse and, therefore, it is not necessary to calculate the dose in them.

9.5.2 CPU time consumption

We ran the four codes on the same machine, an 8-core Intel(R) Xeon(R) CPU X5570 at 2.93GHz, under Scientific Linux 5. The compilation options can have a big effect on the CPU time, as shown in table 9.2, where the results of compiling PENELOPE with two different compilers are given. The version of the codes, the compiler and compiler options used are:

- EGSnrc and DOSXYXnrc: ifort compiler version 10.1 with option –O3.
- PENELOPE: ifort compiler version 10.1 with option –fast and gfortran compiler version 5.2.0 with option –O2.

Table 9.1. Computer memory needed to store the geometry and absorbed dose matrix for the different Monte Carlo packages.

	Geant4/ GAMOS	EGSnrc/ DOSXYZnrc	MCNP6	PENELOPE/ penEasy
Phantom geometry (bytes/voxel)	8	—	4	8
Phantom geometry + dose matrix (bytes/voxel)	152/40	40	98	40

- MCNP: ifort compiler version 10.1 with option –fast.
- GAMOS and Geant4: gcc compiler version 5.2.0 with no optimization options, as they do not represent an advantage in the machine used.

The initialization times and the CPU time per event are shown in table 9.2. It should be noted, that for MCNP we code the time shown as the one used to calculate the dose in organs, while for the other three codes, we quoted the time to calculate the full 3D dose map.

9.5.3 Dose and absorbed fraction scoring

As some of the Monte Carlo codes under study cannot create a uniformly distributed source in a given organ in the phantom, calculate the absorbed dose

Table 9.2. Comparison of the CPU times spent by each of the Monte Carlo codes initializing and processing the events.

	EGSnrc/ DOSXYZnrc	GAMOS/ Geant4	MCNP6	PENELOPE/ PenEasy (ifort –fast)	PENELOPE/ PenEasy (gfortran –O2)
Initialization time (s)	39	95	22,620	97	491
Time per event (μs)	66	126	868 (only organ doses)	28	42

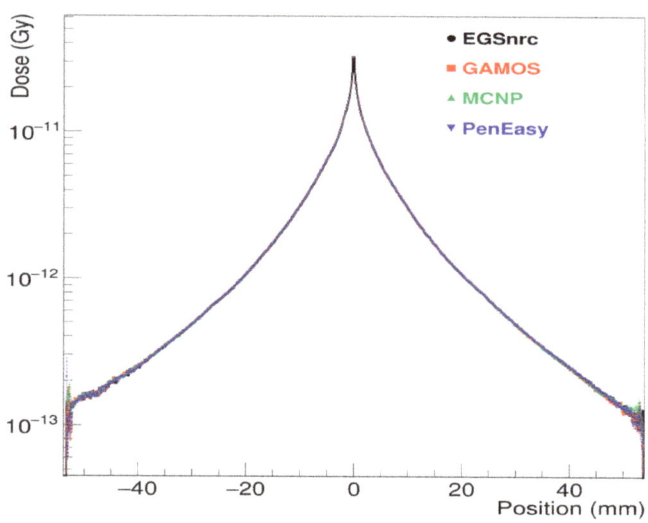

Figure 9.2. Comparison of the calculation of four Monte Carlo codes of the distribution of dose integrated in each of the longitudinal planes of the MOBY phantom calculated with the four Monte Carlo codes. The doses have been normalized to unity in the whole phantom.

per organ, or simulate the full decay chain of an isotope, we have chosen a simple benchmark. The source is a 150 keV gamma placed at the center of the phantom, that is, a position (0,0,0), with a random distribution of direction. We ran 1×10^9 initial particles so that the average dose uncertainty for the voxels that have a dose bigger than 50% of the maximum dose (we take the definition of dose uncertainty as given by APPM Report 105 [53]) is 0.2%. The calculation of a full dose map with this precision serves to compare, in detail, the performance of different codes, but it is probably not needed if the user is only interested in calculating the organ doses. Nevertheless, MCNP6 is not able to compute a dose map ('mesh tally') in so many voxels in the machine we used, but only $128 \times 128 \times 744 = 12\ 189\ 646$ voxels. Therefore, for this code, we have only calculated the dose in organs and the dose in each longitudinal plane.

We compared, in detail, several dose distributions from each code (one- and two-dimensional doses in the *XY*, *XZ*, and *YZ* plane). As an example, we show in figure 9.2 the distribution of dose along the longitudinal axis, integrating the other

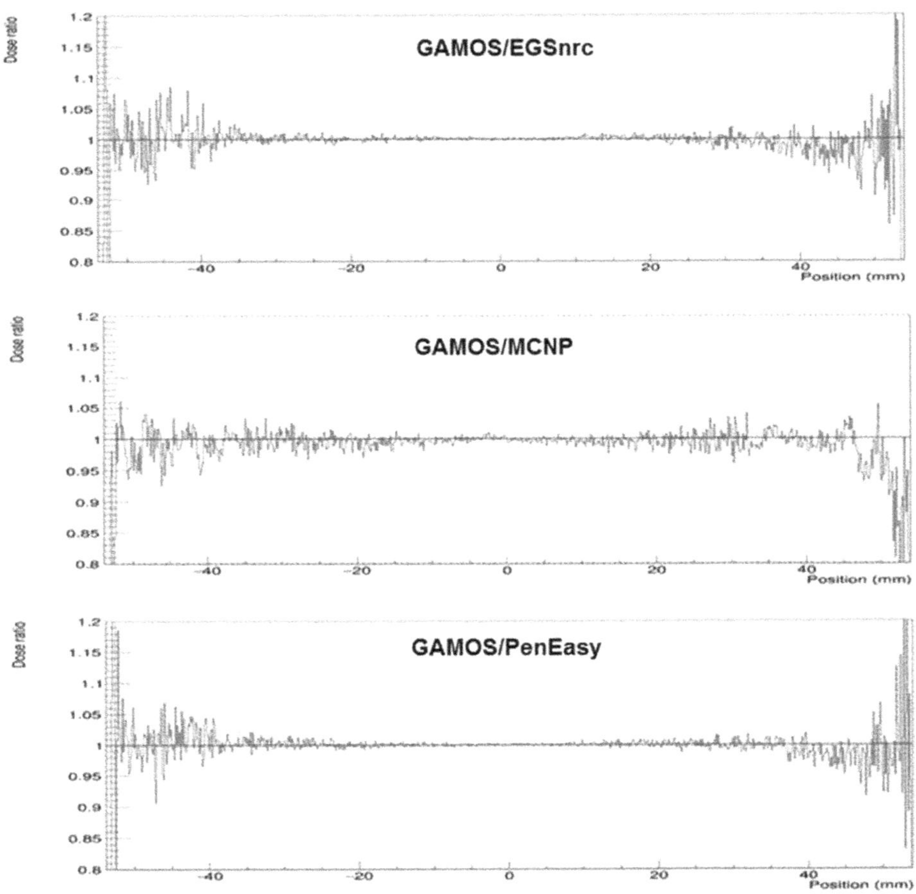

Figure 9.3. Ratio of the doses along the longitudinal axis: GAMOS dose divided by EGSnrc, MCNP and PenEasy dose.

two axis. Other dose map figures show similar behavior. We observe that the distributions are very similar, although there are some differences that are hardly compatible with statistical fluctuations, as can be seen in figure 9.3, where the ratio of the dose obtained with GAMOS to the one obtained with the other three codes is plotted as a function of the position on the longitudinal axis.

The choice of physics options that the different codes propose may have a big impact on the results. We ran GAMOS with the three different packages of electromagnetic physics models and the differences are within statistical errors. We ran EGSnrc/DOSXYZnrc using the cross sections calculated with XCOM, and in this case, there are big differences in the dose far from the center, as can be seen in figure 9.4

Finally, we processed the dose maps to check how the differences in the dose distributions translate into organ AFs, and the results are shown in table 9.3. The errors in all cases are in the fourth or fifth decimal place, so the observed differences are not statistical, but they correspond to differences in the physics models. Figure 9.5 shows the ratio of the organ doses in a graph; the doses calculated by each of the codes are divided by the average of the four codes.

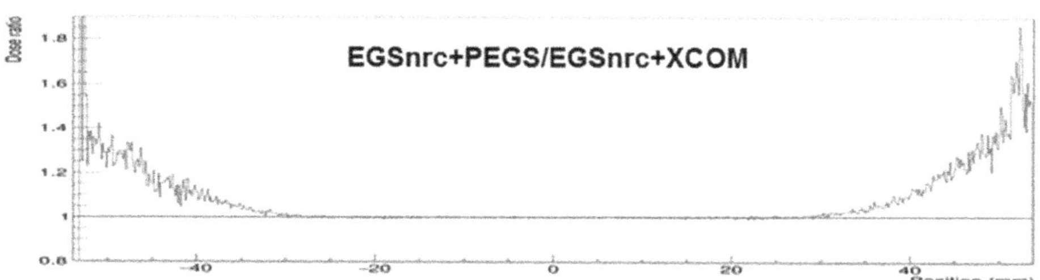

Figure 9.4. Ratio of the doses along the longitudinal axis: EGSnrc with PEGS libraries divided by EGSnrc with XCOM libraries.

Table 9.3. Comparison of the absorbed fractions in several of the MOBY organs calculated with the four Monte Carlo codes (AF are given in %).

ORGAN	EGSnrc/DOSXYZnrc	GAMOS/Geant4	MCNP6	PENELOPE/PenEasy
Lung	0.1429	0.1458	0.1471	0.1454
Intestine	0.2830	0.2835	0.2887	0.2828
Pancreas	0.7141	0.7156	0.7205	0.7135
Brain	0.008 43	0.008 36	0.008 24	0.008 42
Muscle	2.3763	2.3694	2.3448	2.3752
Kidneys	0.009 30	0.009 33	0.009 34	0.009 31
Liver	0.7882	0.7890	0.7939	0.7875
Cartilage	0.0324	0.0325	0.0312	0.0325
Skull	0.0485	0.0489	0.0532	0.0484
Rib bone	0.0719	0.0725	0.0674	0.0716

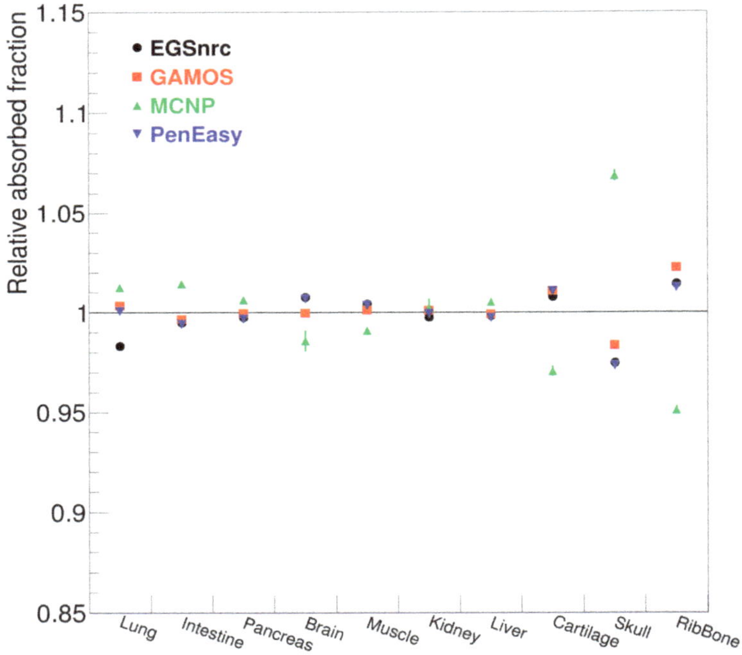

Figure 9.5. Ratio of the organ doses of each of the four Monte Carlo codes divided by the average of the four codes.

9.5.4 Summary

In summary, the four codes behave quite differently in what concerns memory and CPU time consumption but calculate similar distributions of deposited dose. There are, however, some differences in the physics that translate, in most cases, in differences of the order of a few per cent for the calculations of organ AFs not compatible with statistical fluctuations. We have not analysed the ease of use of each code, as this is a subjective matter. Nevertheless, this aspect should also be evaluated before selecting a code, together with the provision of some utilities, like the possibility to calculate the dose or absorbed fraction in an organ directly, the use of an organ as a source of primary particle, the simulation of an isotope with its decay chain, the visualization of the results, or other tools that may help the user to obtain the desired results as well as understanding them.

References

[1] Mauxion T, Barbet J, Suhard J, Pouget J-P and Poirot M 2013 Improved realism of hybrid mouse models may not be sufficient to generate reference dosimetric data *Med. Phys.* **40** 052501

[2] Pennes H H 1948 Analysis of tissue and arterial blood temperature in resting forearm *J. Appl. Physiol* **1** 93–122

[3] Trakic A, Crozier S and Liu F 2004 Numerical modelling of thermal effects in rats due to high-field magnetic resonance imaging (0.5-1 GHz) *Phys. Med. Biol.* **49** 5547–58

[4] Berger M J 1963 Monte Carlo calculation of the penetration and diffusion of fast charged particles *Methods in Computational Physics* ed B Alder, S Fernbach and M Rotenberg (New York and London: Academic Press)
[5] Molière G 1947 Theorie der Streuung schneller geladener Teilchen. Einzelstreuung am abgeschirmten Coulumb-Feld *Z. Naturforsch* **2a** 133–45
[6] Molière G 1948 Theorie der Streuung schneller geladener Teilchen I II. Mehrfach- und Vielfachstreuung *Z. Naturforsch* **3a** 78–97
[7] Molière G 1955 Theorie der Streuung schneller geladener Teilchen III. Die Vielfachstreuung von Bahnspuren unter Berücksichtigung der statistischen Kopplung *Z. Naturforsch* **10a** 177–211
[8] Goudsmit S and Saunderson J L 1940 Multiple scattering of electrons *Phys. Rev.* **57** 24–9
[9] Schneider U, Pedroni E and Lomax A 1996 The calibration of CT Hounsfield units for radiotherapy treatment planning *Phys. Med. Biol.* **41** 111–24
[10] du Plessis F C P, Willemse C A and Lötter M G 1998 The indirect use of CT numbers to establish material properties needed for Monte Carlo calculation of dose distributions in patients *Med. Phys.* **25** 1195–201
[11] Yeom S, Jeong J H, Han M C and Kim C H 2014 Tetrahedral-mesh-based computational human phantom for fast Monte Carlo dose calculations *Phys. Med. Biol.* **59** 3173–85
[12] Nelson W R, Hirayama H and Rogers D W O 1985 The EGS4 Code System, SLAC-265
[13] Kawrakow I and Rogers D W O 2000 The EGSnrc Code System. NRC Report PIRS-701
[14] Hirayama H, Namito Y, Bielajew A F, Wilderman S J and Nelson W R 2005 The EGS5 Code System SLAC-R-730
[15] Böhlen T T, Cerutti F, Chin M P W, Fassò A, Ferrari A, Ortega P G, Mairani A, Sala P R, Smirnov G and Vlachoudis V 2014 The FLUKA Code: Developments and challenges for high energy and medical applications *Nucl. Data Sheets* **120** 211–4
[16] Ferrari A, Sala P R, Fassò A and Ranft J 2005 FLUKA: A multi-particle transport code. CERN-2005-10 INFN/TC_05/11, SLAC-R-773
[17] Allison J *et al* 2016 Recent developments in Geant4 *Nucl. Instrum. Methods Phys. Res.* A **835** 186–225
[18] Allison J *et al* 2006 Geant4 developments and applications *IEEE Trans. Nucl. Sci* **53**(1) 270–8
[19] Agostinelli S *et al* 2003 Geant4 - A Simulation Toolkit *Nucl. Instrum. Methods Phys. Res.* A **506** 250–303
[20] 2003 X-5 Monte Carlo Team. MCNP — A General Monte Carlo N-Particle Transport Code, Version 5. Volume I: Overview and Theory. Report LA-UR-03-1987
[21] Pelowitz D B 2011 MCNPX User's manual Version 2.7. Report LA-CP-11-00438
[22] Werner C J 2017 MCNP User's manual Code Version 6.2. Report LA-UR-29981
[23] Baro J, Sempau J, Acosta E, Fernández-Varea J M and Salvat F 1995 Penelope, an algorithm for Monte Carlo simulation of the penetration and energy loss of electrons and positrons in matter *Nucl. Instrum. Methods Phys. Res.* B **100**(1) 31–46
[24] Sempau J, Acosta E, Baro J, Fernández-Varea J M and Salvat F 1997 An algorithm for Monte Carlo simulation of coupled electron-photon transport *Nucl. Instrum. Methods Phys. Res.* B **132** 377–90
[25] Salvat F PENELOPE-2014: A Code System for Monte Carlo Simulation of Electron and Photon Transport. OECD Nuclear Energy Agency 2015. Available in pdf format at http://www.oecd-nea.org

[26] Rodriguez M, Sempau J and Brualla L 2013 PRIMO: A graphical environment for the Monte Carlo simulation of Varian and Elekta linacs *Strahlenther. Onkol.* **189** 881–6
[27] Arce P *et al* 2014 Gamos: A framework to do Geant4 simulations in different physics fields with a user-friendly interface *Nuc. Instr. Methods Phys. Res.* A **735** 304–13
[28] Jan S *et al* 2004 GATE - Geant4 Application for Tomographic Emission: a simulation toolkit for PET and SPECT *Phys. Med. Biol.* **49** 4543–61
[29] Perl J, Shin J, Schümann J, Faddegon B and Paganetti H 2012 TOPAS: An innovative proton Monte Carlo platform for research and clinical applications *Med. Phys.* **39** 6818–37
[30] España S, Herraiz J L, Vicente E, Vaquero J J, Desco M and Udias J M 2009 PeneloPET, a Monte Carlo PET simulation tool based on PENELOPE: Features and validation *Phys. Med. Biol.* **54** 1723–42
[31] Zerby C D and Moran H S 1962 *A Monte Carlo calculation of the three-dimensional development of high-energy electron-photon cascade showers* (Oak Ridge, TN: Oak Ridge National Laboratory) Report ORNL-TM-422
[32] Zerby C D and Moran H S 1962 *Studies of the longitudinal development of high-energy electron photon cascade showers in copper* (Oak Ridge, TN: Oak Ridge National Laboratory) Report ORNL-3329
[33] Zerby C D and Moran H S 1963 Moran. Studies of the longitudinal development of electron-photon cascade showers *J. Appl. Phys.* **34** 2445–57
[34] Nagel H H 1964 Die Berechnung von Elektron-Photon-Kaskaden in Blei mit Hilfe der Monte-Carlo Methode. Inaugural-Dissertation zur Erlangung des Doktorgrades der Hohen Mathematich-Naturwissenschaftlichen Fakultät der Rheinischen Freidrich-Wilhelms-Universität, Bonn
[35] Nagel H H 1965 Elektron-Photon-Kaskaden in Blei: Monte-Carlo-Rechnungen für Primärelektronenergien zwischen 100 und 1000 MeV *Z. Physik* **186** 319–46
[36] Nagel H H and Schlier C 1963 Berechnung von Elektron-Photon-Kaskaden in Blei fur eine Primärenergie von 200 MeV *Z. Physik* **174** 464–71
[37] Kawrakow I, Mainegra-Hing E, Rogers D W O, Tessier F and Walters B R B 2017 The EGSnrc Code System: Monte Carlo simulation of electron and photon transport *Technical Report PIRS-701* (Canada: National Research Council)
[38] Ivanchenko V N, Maire M and Urban L 2004 Geant4 standard electromagnetic package for HEP applications *Proc. Conf. Rec.* (Rome, Italy: IEEE Nuclear Science Symp) N33–179
[39] Cullen D, Hubbell J H and Kissel L 1997 EPDL97: the Evaluated Photon Data Library, 97 Version *Lawrence Livermore National Laboratory, Rep.* UCRL-50400 **vol. 6**
[40] Perkins S T, Cullen D E and Seltzer S M 1997 Tables and Graphs of Electron-Interaction Cross-Sections From 10 eV to 100 GeV Derived From the LLNL Evaluated Electron Data Library (EEDL), Z=1 –100 *Lawrence Livermore National Laboratory, Rep.* UCRL-50 400 **vol. 31**
[41] Perkins S T, Cullen D E, Chen M H, Hubbell J H and Rathkopf J 1991 Scofield J. Tables and Graphs of Atomic Subshell and Relaxation Data Derived From the LLNL Evaluated Atomic Data Library (EADL), Z=1–100 *Lawrence Livermore National Laboratory, Rep.* UCRL-50 400 **vol. 30**
[42] Amako K *et al* 2005 Comparison of Geant4 Electromagnetic Physics Models Against the NIST Reference Data *IEEE Trans. Nucl. Sci.* **52**(4) 910–8
[43] Geant4 Collaboration Geant4 Physics Reference Manual. http://geant4.web.cern.ch/geant4/UserDocumentation/UsersGuides/PhysicsReferenceManual/html/index.html

[44] Arce P, Apostolakis J and Cosmo G 2008 A technique for optimised navigation in regular geometries *Nuclear Science Symp. Conf. Record* NSS '08. IEEE

[45] Hughes H G 2014 Recent developments in low-energy electron/photon transport for MCNP6. Prog *Nucl. Sci. Technol.* **4** 454–8

[46] Sempau J, Badal A and Brualla L 2011 A PENELOPE-based system for the automated Monte Carlo simulation of clinacs and voxelized geometries--application to far-from-axis fields *Med. Phys.* **38** 5887–95

[47] Berger M J, Hubbell J H, Seltzer S M, Chang J, Coursey J S, Sukumar R and Zucker D S 2005 *XCOM: Photon Cross sections Database* (Gaithersburg, MD: Technical report, Institute of Standards and Technology) Numerical tables available at http://physics.nist.gov/PhysRefData/Xcom/Text/XCOM.html

[48] ICRU Report 77 2007 *Elastic Scattering of Electrons and Positrons* (Bethesda, MD: ICRU)

[49] Salvat F, Jablonski A and Powell C J 2005 ELSEPA--Dirac partial-wave calculation of elastic scattering of electrons and positrons by atoms, positive ions and molecules *Comput. Phys. Commun.* **165** 157–90

[50] García-Toraño E, Peyres V, Bé M M, Dulieu V, Lépy M C and Salvat F 2017 Simulation of decay processes and radiation transport times in radioactivity measurements *Nucl. Instrum. Methods Phys. Res.* B **396** 43–9

[51] Hughes H G 2015 An Electron/Photon/Relaxation Data Library for MCNP6. Report LA-UR-13-27377

[52] Segars W P, Tsui B M, Frey E C, Johnson G A and Berr S S 2004 Development of a 4-D digital mouse phantom for molecular imaging research *Mol. Imaging Biol.* **6** 149–59

[53] Chetty I J *et al* 2007 Report of the AAPM Task Group No. 105: Issues associated with clinical implementation of Monte Carlo-based photon and electron external beam treatment planning *Med. Phys.* **34** 4818–53

Part II

Applications in preclinical research

Chapter 10

Applications of computational animal models in ionizing radiation dosimetry

Panagiotis Papadimitroulas and George C Kagadis

10.1 Introduction

Modern medicine is extremely interested in the quantification of absorbed doses at organ/tissue level considering both the anatomical and physical variabilities of the specimen. In ionizing radiation, the assessment of energy deposition is a crucial parameter for the analysis of the risks of cancer and the evaluation of clinical and preclinical protocols. Over the last few years, a big effort has been put into combining radiation dosimetry and biological effects at the cellular and molecular levels. The introduction of computer science in medical physics is continuing to evolve, overcoming several drawbacks of the past. Laboratory animals are widely used in experimental procedures, to calculate and to define dosimetry at organ level, or even to extract radiobiological effects from internal and external ionizing radiation. This effort aims to translate the research findings 'from mouse to man' evaluating the preclinical and clinical protocols that involve ionizing radiation [1].

Monte Carlo simulations have become a standard tool for ionizing radiation applications, with robust and fast calculations of complex experiments [2]. Speed-up techniques (high performance computers—HPC, variance reduction techniques—VRTs, graphical processing units—GPUs and clusters) have been developed to execute realistic simulations. Alongside, the development of computerized models has increased the efficiency of accurate modeling of the anatomy of the specimens. Over the last decade there has been a great increase in the different small animal models (mouse, rat, canine, frog, fish, etc see Part I of the book) that were developed with more and more accuracy of the description of the anatomical characteristics.

The development of such tools allowed the increased accuracy of the absorbed dose assessment in several preclinical applications using ionizing radiation. A variety of popular Monte Carlo (MC) codes have been used (Geant4 [3], MCNP [4], EGS [5], etc), promoting them as the 'gold standard' for radiation dosimetry. The high

doi:10.1088/2053-2563/aae1b4ch10

resolution modern computational models, provide accurate energy deposition in voxel and even at molecular level, allowing the definition of energy deposition in several applications, such as internal radioimmunotherapy (RIT), external radiation therapy (RT), internal dosimetry using radiotracers, dosimetry through imaging procedures (PET, SPECT, CT) and the micro- and nano-dosimetry in cellular and molecular level (G4-DNA [6]). Great efforts have been made over the years in the dosimetry assessment of small animal applications, developing dosimetry databases of radiotracers, evaluating and validating the absorbed fractions (AFs), the specific absorbed fractions (SAFs), the *S*-values as they are defined in medical internal radiation dose (MIRD) schema [7, 8]. Biokinetics of the tracers, are taken into account so as to further evaluate the dosimetry, and several parameters affecting absorbed dose has also been extensively studied, including the age of the animal, the size, mass, location, and density of the organs/tissues. The use of MC simulations also provided the ability to study the dose dependency on the decay scheme of several radionuclides, the type, energy, and yield of the emission and accurately define the uncertainty in voxel level [9].

According to 'Scopus' [10], after 2000, there was a large growth in the number of scientific studies using MC simulations and computational animal models for radiation dosimetry applications. The development of numerous animal phantoms in combination with the evolution of fast MC techniques increased the applications in the investigation of dose assessment in preclinical medicine. Figure 10.1 depicts the trend of the increasing studies in the field since the 1990s.

10.2 Fundamentals of radiation dosimetry

A definition of absorbed dose has been given by the International Commission on Radiation Units and Measurements [11]:

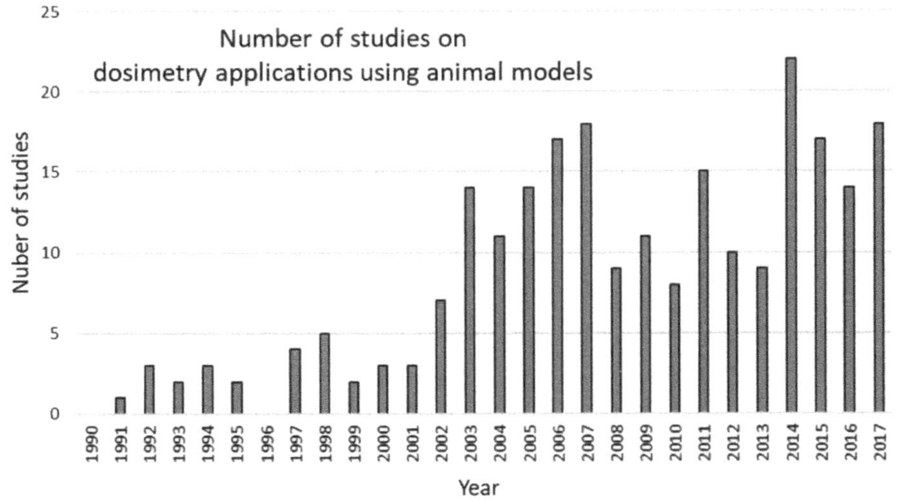

Figure 10.1. Number of dosimetry studies using MC simulations and computational animal models since the 1990s in 'Scopus'.

$$D = \frac{d_{\bar{\varepsilon}}}{dm} \tag{10.1}$$

This represents the mean energy $d_{\bar{\varepsilon}}$ imparted by ionizing radiation to matter of mass dm. The SI unit for absorbed dose is the gray (Gy), referring to energy of 1 J absorbed in 1 kg of matter (J kg^{-1}). For humans the quantity 'equivalent dose', is defined as $H_{T,\,R} = w_R D_{T,\,R}$. The quantity $H_{T,\,R}$ is the equivalent dose for a specific tissue T and radiation type R, while w_R is the radiation weighting factor, associated with the relative biological effectiveness (RBE). It expresses the ratio of doses of different radiation types needed to produce a given biological endpoint (e.g. 63% cell killing in an *in vitro* cell survival study).

Another defined quantity for humans is the 'effective dose'. This is the tissue-weighted sum of the equivalent doses to all specified tissues of the body [12]. H_T is the equivalent dose in a tissue or organ, while w_T is the corresponding weighting factor for that specific organ.

The sum of the weighted equivalent doses for a given exposure is the 'effective dose' (E):

$$E = \sum\nolimits_T w_T \, H_T \tag{10.2}$$

The effective dose is intended to represent the equivalent dose that, if received uniformly by the whole body, would result in the same total risk as that actually incurred by a given actual non-uniform irradiation. The unit for this quantity is the same as for absorbed dose (J kg^{-1}), and its special name is Sievert (Sv).

The use of effective dose helps to simplify the way we report dose in the cases where different doses are received by different organs. This way we can directly compare different radiological examinations, e.g. a computed tomography and a nuclear medicine scan. The corresponding weighting factors incorporate all the differences of the various radiations used in every situation. Furthermore, effective dose should only be used when discussing risks to populations, and it should be made clear that it should also only be used when discussing doses to humans.

10.2.1 Nuclear medicine dosimetry

The calculation of absorbed dose to an organ from an internally distributed radionuclide has been described in detail by the medical internal radiation dose (MIRD) committee (Society for Nuclear Medicine and Molecular Imaging—SNMMI). In a simple form, the MIRD method for calculations of absorbed dose requires the total number of nuclear transitions over a specific time, as well as the amount of deposited energy per nuclear transition and unit mass of the target organ [13]. In nuclear medicine we are interested in the estimation of the absorbed dose in a given organ/tissue from a specific radiopharmaceutical administration. Thus, a generic equation for the absorbed dose rate in an organ is calculated by the average amount of deposited energy per unit mass, as shown in equation (10.3)

$$\dot{D} = \frac{kA \sum_i n_i \, E_i \, \varphi_i}{m} \tag{10.3}$$

$\dot{D}$ is the absorbed dose rate in Gy s^{-1}, k is a proportional constant, A is the activity in MBq, n is the number of nuclear transitions with energy E, E is the energy per irradiation (i) in MeV, φ is the energy fraction emitted from a source and m is the mass of the target organ.

In the MIRD formalism [14], the absorbed dose $D(r_T)$ in any target organ (r_T) from a source organ (r_S) is estimated by the equation:

$$D(r_T) = \sum_{r_S} S(r_T \leftarrow r_S)\tilde{A}(r_S) \tag{10.4}$$

Summing over all source organs (r_S), the products of the time-integrated activity $\tilde{A}(r_S)$ in each source organ with the corresponding S-value (S). The S-value is the mean absorbed dose to the target region (r_T) per unit of nuclear transition of the relevant radionuclide in the source region considered. The S-value is defined by the equation:

$$S(r_T \leftarrow r_S) = \frac{1}{M(r_T)} \sum_i E_i Y_i \varphi(r_T \leftarrow r_S, E_i) \tag{10.5}$$

where $\varphi(r_T \leftarrow r_S)$ is the absorbed fraction (AF) at energy E_i, which is defined as the fraction of radiation E_i emitted within the source organ r_S, that is absorbed in a target organ r_T. Y_i is the radiation yield of the ith nuclear transition per nuclear transformation and $M(r_T)$ is the mass of the target organ. Finally, the specific absorbed fraction (SAFs) are derived from the AFs as shown in equation (10.6):

$$\Phi(r_T \leftarrow r_S, E_i) = \frac{\varphi(r_T \leftarrow r_S, E_i)}{M(r_T)} \tag{10.6}$$

Nevertheless, appropriate information or assumptions are required on the

- activity of the administered radiopharmaceutical,
- radionuclide decay rate (physical half-life),
- fraction of energy and type per decay (spectrum information),
- fraction of the administered activity present in each source organ (time activity function, and uptake),
- timeframe the activity is present in the source organ (effective half-life),
- total number of decays in the source organ (cumulated activity),
- fraction of energy absorbed in the target organ from radioactivity present in the source organ (absorbed fractions), and
- mass of each specific organ.

The absorbed dose is defined as the mean imparted energy $d\bar{\varepsilon}$ to matter of mass dm and is expressed with the unit Gy [J kg^{-1}] as indicated in equation (10.1) above.

10.2.2 Computed tomography (CT) dosimetry

X-ray micro-computed tomography (μCT) imposes a moderately high ionizing radiation dose on the animal, since larger photon fluence is necessary to sustain the signal-to-noise ratio high enough at high spatial resolution, so that the image quality is adequate. Although the dose level is far from being fatal, it might be high enough to change physiological parameters in the animal that could confound the biological outcome of the experiment [15–17]. Studies in the scientific literature investigating the biological effects of radiation doses relevant to μCT have been reviewed by Meganck *et al* [17].

Therefore, it is essential to have an accurate understanding of radiation dose within a scan. Unfortunately, there is not a single way to describe the x-ray dose. Literature studies have used Roentgens (R), Sieverts (Sv), Gray (Gy), radiation absorbed dose (rad), and Roentgen equivalent man (REM) [15, 16]. In clinical environments, three types of dosimetric quantities are used in CT dosimetry [18–20]. Also, there is no analogous standard method for preclinical research. Few techniques have been applied in the current literature. The simplest and most widely used technique for CT dosimetry is the use of an ionization chamber [21–27].

CT dose index ($CTDI_{100}$) is calculated as defined in units of mGy using equation (10.7) below [28]:

$$\mathrm{CTDI}_{100} = \frac{1}{\mathrm{NT}} \int_{-50\ \mathrm{mm}}^{50\ \mathrm{mm}} D(z)\, dz \tag{10.7}$$

where $D(z)$ is the dose profile along the z-axis, N is defined as the number of slices, and T is defined as the slice thickness at the isocenter. $N * T$ is effectively the beam width in the z-direction at the isocenter. The f-factor of 0.87 rad R^{-1}, noted for 120 kVp [28], is also induced to convert the amount of ionization in air (roentgens or coulombs kg^{-1}) to absorbed dose in tissue (rads or grays). $CTDI_{100}$ represents the accumulated multiple scan dose at the center of a 100 mm scan and underestimates the accumulated dose for longer scan length [29, 30]. The CTDI varies across the field of view (FOV). The average CTDI across the FOV is estimated by the Weighted CTDI ($CTDI_w$) [29] as indicated below:

$$\mathrm{CTDI}_w = 1/3*\mathrm{CTDI}_{100,\mathrm{center}} + 2/3*\mathrm{CTDI}_{100,\mathrm{edge}} \tag{10.8}$$

$CTDI_{vol}$ is a useful indicator of the dose to a standard phantom in a specific exam protocol. It considers protocol-specific information like the pitch.

$$\mathrm{CTDI}_{\mathrm{vol}} = \frac{N*T}{I}*\mathrm{CTDI}_w \tag{10.9}$$

where the pitch is defined as the ratio of the table movement per rotation (I), to the total nominal beam width ($N * T$).

The absorbed dose can be integrated along the scan length to compute the dose–length product (DLP) to estimate the overall energy delivered by a given scan protocol [31].

$$\mathrm{DLP(mGy*cm)} = \mathrm{CTDI}_{\mathrm{vol}}*\mathrm{scan\ length(cm)} \tag{10.10}$$

The DLP reveals the total energy absorbed (and thus the potential biological effect) which may be attributed to the complete scan acquisition. It is fundamental to recognize that the potential biological effects from radiation not only on the radiation dose to a specific tissue or organ. They are also affected by the biological sensitivity of the tissue or organ irradiated. The effective dose, *E*, is a dose descriptor that reveals this difference in biologic sensitivity [32] and, as already mentioned, its unit is the Sievert. It is a single dose parameter that reflects the risk of a non-uniform exposure in terms of an equivalent whole-body exposure.

One direct way of estimating doses to patients undergoing CT examinations is to measure organ doses in animal-like phantoms, with thermo-luminescence dosimeters (TLDs) or nanoDots placed near specific organs within an animal phantom to measure the doses at specific locations, or with gafchromic films [26, 33–36].

The most accurate way of obtaining the pattern of energy deposition in small animals undergoing CT examinations is by Monte Carlo simulations in computational anatomical animal models. Many studies have been carried out to estimate the radiation doses of small animals using Monte Carlo modeling code [26, 33, 37–43]. A study by Boone *et al* [37] provides tables to calculate whole-body x-ray dose estimates from microCT to a cylindrical mouse of homogeneous media. Jones *et al* [39] used Monte Carlo techniques to calculate the mean doses received by 20 organs during diagnostic x-ray examinations for 45 combinations of tube voltage and filtration ranging from 50 to 140 kVp and 1.5–4 mm of aluminum, respectively. Lee *et al* [40] evaluated the absorbed doses of 37 brain tissues and major organs of the mouse according to kVp changes with the MOBY phantom and measured data using TLDs. Lately, GATE Monte Carlo simulations were used by Taschereau *et al* [41] to calculate the radiation dose received by individual organs of a mouse subjected to microCT imaging procedures with the MOBY phantom.

10.2.3 Multimodality (SPECT/CT and PET/CT) dosimetry

The use of 'hybrid imaging' using SPECT/CT and PET/CT has grown significantly over the past few years, increasing concerns for the total radiation dose per examination. In hybrid examinations, regardless of the acquisition protocol, the effective doses from external and internal sources are first estimated separately in different ways, and are then combined to give a total whole-body radiation dose. Effective dose is an indicative quantity that shows the differences of the radiation dose from the radiopharmaceutical and the x-ray part of a hybrid examination, evaluating the relative biological risks [44]. Low doses of ionizing radiation related to medical imaging examinations, particularly CT, are not dangerous according to radiation protection and medical physics organizations. Nevertheless, imaging professionals focus on the ALARA (as low as reasonably achievable) principle, and dose management as there is a lack of documentation of the long-term effects in the low-dose range and risk estimation below 100 mSv have huge uncertainties [45–49]. For these reasons, the interest in keeping the radiation dose as low as is clinically practical is increased. Hybrid examinations introduce an extra complexity, especially regarding the CT scan. In the case where a recently acquired CT scan is available, the CT may be used only for anatomic correlation of the PET acquisition rather than directly for diagnosis. Therefore, the level of radiation dose may vary a lot, depending on its use [50–53]. Regarding the SPECT/CT scanners, some of the systems have standard CT scanners, whereas others have CT parts only for attenuation correction and anatomical localization of the uptake on SPECT and are unable to produce diagnostic quality images. The radiation dose from each approach of the CT scanner may be high. In order to achieve dose reduction, before the SPECT/CT is acquired, a planar whole-body scan is often pre-acquired [52, 54].

10.3 Applications in ionizing radiation dosimetry

Over the last few decades, science has evolved rapidly in dosimetry assessment due to ionizing radiation. Preclinical studies are extensively used for the standardization of dosimetry protocols and the evaluation of diagnostic tools and therapeutic schemes. MC simulations provide all the appropriate tools for accurate computations in terms of physical processes, particle tracking and energy deposition. The involvement of computational animal models in MC calculations has increased the accuracy of absorbed dose assessment in internal and external radiation procedures. A great deal of effort is put into the accurate anatomical modeling of several animals that are extensively used in preclinical dosimetry experiments.

Modern science trends are towards individualized protocols both for diagnosis and dosimetry applications. The combination of the MC calculations and the computational animal models, provide the estimation of absorbed dose taking into account the inhomogeneity within the body and determine the amount of the deposited energy in various organs/tissues by external or internal ionizing sources. Thus, the computational models serve to mimick the interior and exterior anatomical features of a human or animal body.

Computational phantoms were integrated with MC simulations for the first time in the 1960s and were used to simulate the radiation transport and the particle interactions within the body. The rapid evolution in genetics and molecular imaging, combined with the development of techniques for genetically engineering animals, increased the development of a large list of animal computational phantoms, as they are described in Part I of the current volume. Figure 10.2 shows an indicative example of a computational mouse model used in an internal radiation dosimetry simulation. The biodistribution of ^{68}Ga-NODAGA-RGD [55] was used in the MOBY phantom with its respective attenuation, and dosimetry map which resulted from the GATE MC toolkit.

10.3.1 Monte Carlo simulations

Computational phantom advances with the rapid development of computing and processing tools, have increased applications in preclinical radiation studies. Monte Carlo calculations have been considered to be the reference tool for the evaluation and validation of new methods and algorithms involving the use of ionizing radiation in diagnostic and therapeutic medical physics. Furthermore, they have been considered the ‘gold standard’ technique for radiation dosimetry calculations in preclinical research.

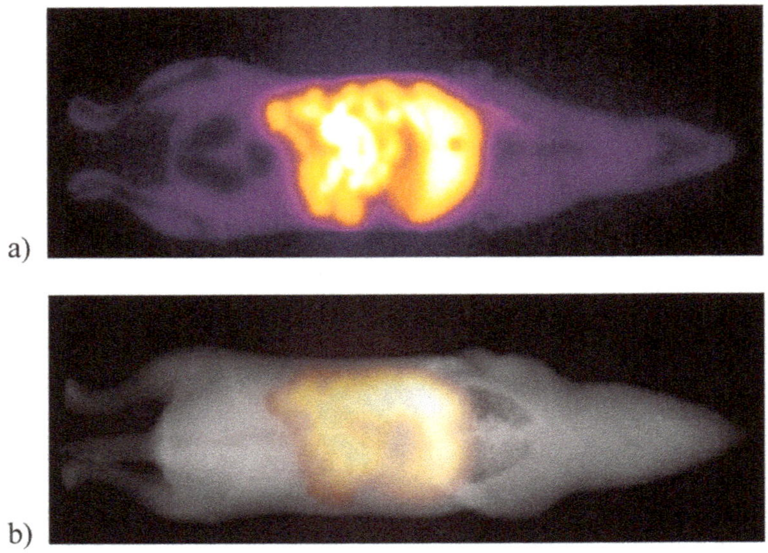

Figure 10.2. Simulation of the MOBY mouse model. (a) Whole-body view of a central slice of the dosimetry map generated by GATE, and (b) merged attenuation and dosimetry map of the mouse model.

MC simulations are statistical approaches based on the use of random number generators to converge on the solution of a specific deterministic problem. Therefore, the use of such simulations is necessary to approach a solution of the underlying physical processes that are taking place in ionizing radiation interactions. A critical aspect of MC simulations in the field of radiation dosimetry is the physical interactions efficiency calculation, incorporating attenuation probabilities, low energy photon interactions, particle multiple-scattering, electron ionization, production of secondary charged particles, etc. Preclinical radiation studies have increased rapidly with the combined use of MC simulation and the computational animal models [55]. Figure 10.3 represents an indicative photon irradiation example of the MOBY mouse model which is inserted in the GATE MC toolkit [2].

10.3.2 Dosimetry applications in mouse models

For a few years, stylized computational animal models were used for radiation dosimetry studies. The replacement of these phantoms arrived by the use of CT image data for the creation of realistic whole-body models [56]. Research in the field of small animal diagnosis and therapy was based on simplified models using geometrical shapes, for individual organs.

Since the 1990s, computational animal models started to be developed for dosimetry purposes by Hui *et al* [57]. These models represented a series of 10 athymic mice incorporating a list of 13 major organs. The medical internal radiation dose (MIRD) dosimetry method was extended from humans to mice for beta dose calculations of Y-90, taking into account the cross-organ absorbed fractions (AFs). This procedure provided, for the first time, a more realistic organ-based dosimetry approach using Berger's point kernels and the EGS4 electron transport MC code. Y-

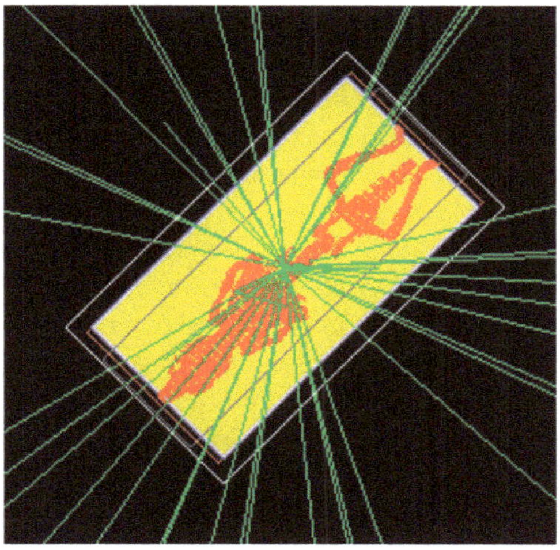

Figure 10.3. A simulation photon irradiation of the MOBY mouse model, in the GATE MC toolkit.

90 was extensively studied for a variety of preclinical applications using mouse and rat models. Kennel *et al* [58] used a 30 g mouse model with approximate geometrical shapes to represent the whole body and several organs in the MCNP-4A code. Treatment of lung tumor radioimmunotherapy (RIT) was tested, comparing the dosimetry assessment of Y-90 and the α-particle emitter Bi-213. The cross-organ energy deposition by β-emitters Y-90 and I-131 were also tested by Flynn *et al* [59] considering the effects of heterogeneity in kidneys and tumors in mice. Stylized mouse models were also used incorporating ellipsoids and/or cylinders for the majority of the organs, while the heterogeneity of the kidneys and tumors was addressed by using two compartments for the cortex and medulla and for the viable and necrotic regions, respectively. This study was an improvement on dosimetry calculations on AFs considering the self-dose and the cross absorbed fractions in heterogeneous regions quantifying the dose differences of necrotic and vital cells within the tumor. Voxelized mouse models were introduced by Kolbert *et al* [60] in dosimetry studies, in the early 2000s. MCNP-4A code, in combination with computational mouse models based on MRI female athymic mouse images, were used to calculate photon and electron *S*-values, for the most important organs, namely; liver, kidneys, spleen. *S*-values for I-131, Sm-153, P-32, Re-188 and Y-90 were calculated and compared assuming the complete electron energy deposition.

In 2004, Funk *et al* [61], mentioned the need for the accurate establishment of absorbed dose in small animal nuclear imaging applications. Thus, the MIRD schema was used for two different stylized models, a mouse and a rat, 30 g and 300 g, respectively, to calculate the *S*-values of a group of radionuclides (F-18, Tc-99m, Tl-201, In-111, I-123 and I-125) used in SPECT and PET acquisitions. The authors concluded that whole-body absorbed doses both in mice and rats, ranges in a few decades of cGys, while there should be a very careful monitoring of administered activity so as to keep absorbed doses lower than the lethal dose to mice, which is considered to be ~7 Gy. To date, the accurate determination of the *S*-values for an extended list of organs in preclinical studies, is of high interest. Hindorf *et al* [62] studied the parameters that influence the calculatiuon of *S*-values both for therapy and imaging radionuclides (Y-90, I-131, In-11, Tc-99m). The organ mass, the geometry, the location, and the shape of the organs play a crucial role in dose assessment, as well as the radiation emission type. The calculation of the beta absorbed fractions and the *S*-values was extended by Miller *et al* [63] for a variety of mouse organs using the MMCNP code and stylized models. The radionuclides tested are Y-90, Re-188, Ho-166, Pm149, Cu-64, and Lu-177. Dosimetry of pure β-emitting Y-90 radionuclide was further investigated to validate the EGS5 code and to calculate the absorbed doses in a tough-water phantom simulating a mouse, with experimental measurements [64].

In 2006 Stabin *et al* [65], introduced AFs and dose factors based on voxelized realistic mouse and rat models. Until then, stylized models were used for assessing dosimetry in preclinical studies. Using the MCNP code and the segmented CT based computational models, AFs were established at discrete energies for both electron and photon sources. One year later, Bitar *et al* [66] introduced a dosimetric database for preclinical targeted radiation therapy experiments, using high resolution voxelized mouse models from digital photographs. The developed database included 13 sources and 25 target organs for 16 photon, and 16 electron energies. In addition, AFs and *S*-values were calculated for 16 radionuclides of interest. A few years later, Keenan *et al* [56] used the non-uniform rational B-spline (NURBS) based MOBY (mouse) and ROBY (rat) model and the Geant4 MC tool to verify their use in SAFs calculation, and in internal and external dosimetry applications. CT dosimetry was also tested in computational voxelized mouse models [67], studying the dependence of the dosimetric characteristics on photon beam energy and tissue heterogeneity. Three mouse models were tested using 100 and 225 kVp photon beams in the DOSXYZnrc code associated with the EGSnrc. Larsson *et al* [68] used the MOBY voxelized mouse model to calculate absorbed doses in tumors using the MCNPX MC code. More specifically, different small half spheres were included in the standard 3D voxel-base model, to represent tumors of several sizes. Dose calculations were performed for both electron and photon monoenergetic sources, and for commonly used radionuclides, concluding that for accurate tumor dosimetry assessment in mouse models it is necessary to precisely consider the activity distribution within the body of the mouse. EGS4 MC code was used incorporating the Digimouse voxel-model to quantify the impact of the voxel size on the calculation of AFs and *S*-values [69]. An in-house voxelized 28 g mouse model was used for the calculation of organ dose conversion coefficients for application using external photon beams [70]. Dosimetry results for photon irradiation from 10 keV up to 10 MeV resulted in the knowledge that organ absorbed doses are sensitive to the body size for energy irradiations up to 0.1 MeV. The use of voxelized phantoms in preclinical dosimetry increased during the last 10 years, for specifying as accurately as possible the influence of the several parameters on absorbed dose. In 2013, Xie and Zaidi [71] generated a dosimetric database of *S*-values for eight of the most common positron emitting radionuclides. They used 17 different mouse models based on the MOBY phantom, so as to study the impact of the body and the organ mass, on the resulting deposited energy. More recently Mauxion *et al* [72] studied the impact of the different mouse models on the dosimetry assessment. The authors used the MCNP and the GATE MC tools to investigate the *S*-values that derived from two 'similar' realistic mouse models, concluding in large variations between the *S*-values. Lately, Kostou *et al* [73] have created a dosimetric database with the *S*-values of eight commonly used radioisotopes in nuclear imaging and therapy. The GATE MC toolkit and the MOBY model were used to study the impact of the absorbed dose when scaling the organ masses. As there is growing interest in the radiological risk of ionizing radiation, dosimetry has also been investigated in small animal CT imaging studies creating a dosimetric database which can be used in targeted radiotherapy experiments [40]. Lately, MC simulations have also been used to investigate the enhancement of absorbed dose when gold nanoparticles (GNPs) are used in the irradiation setup [74]. Figure 10.4 represents, the dose maps (absorbed dose per voxel) of a mouse model after a preclinical x-ray irradiation. The mouse model was used in the simulation, with three different concentrations of GNPs in the liver [75].

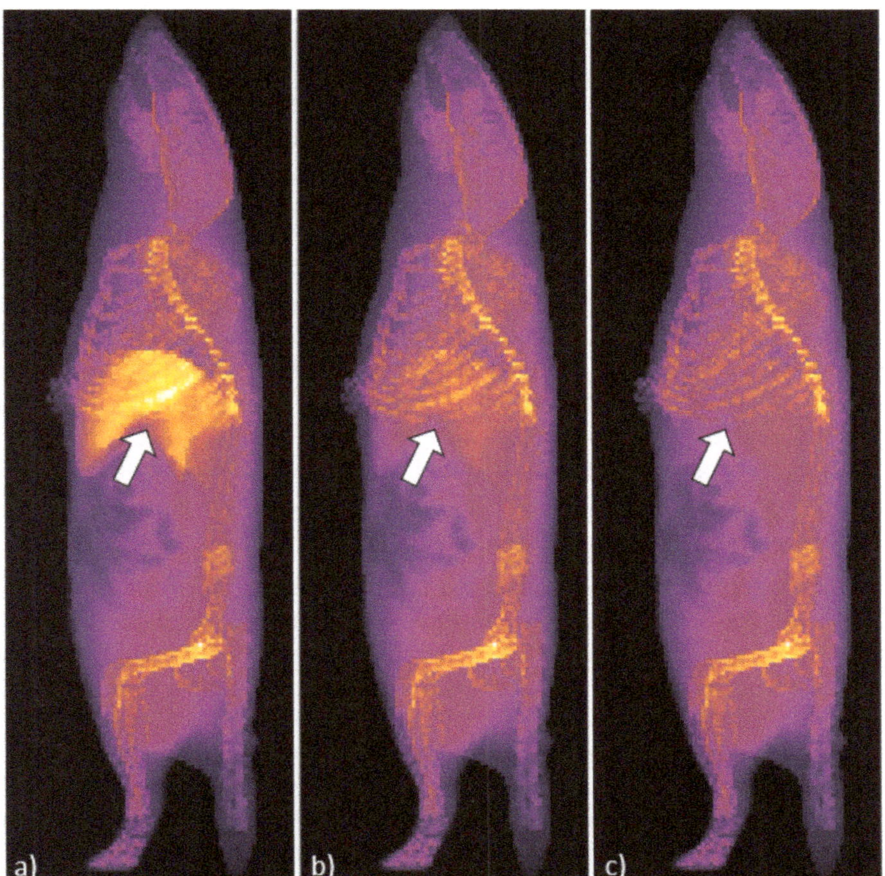

Figure 10.4. Merged slices of simulated dose maps in a mouse model with different concentrations of GNPs in liver: (a) 40 mg Au—0.4% per weight, (b) 10 mg Au—1.64% per weight, and (c) no GNPs present.

10.3.3 Dosimetry applications in rat models

In parallel to the dosimetry applications on mouse models, rat models were extensively used for internal dosimetry irradiations, calculating *S*-values and SAFs [61, 65]. Konijnenberg *et al* [76] used a mathematical model of a rat body in the MCNP code, to evaluate the Y-90, In-11, and Lu-177 dosimetry on radionuclide therapy. The ICRP commission in 2008 [77], introduced the concept of reference animals, describing a 314 g rat model, calculating the dose conversion factors for 75 radionuclides. Similar to the dosimetry simulation in mouse models, Peixoto *et al* [78] used the MCNP code and a Wistar rat model based on CT data for the evaluation of photon and electron AFs. Another realistic, anatomically image-based rat model was developed by Wu *et al* [79] for the assessment of dose conversion coefficient for monoenergetic photon beams in the 10 keV up to 10 MeV range. The absorbed doses of 13 major organs were calculated providing the influence of the anatomical characteristics to the organ dose. Pain *et al* [80] proposed a method using anatomical rat atlases to evaluate and correct the uncertainties in PET radiotracer irradiation dosimetry. Rat voxelized models were also used for specific radionuclide therapy applications in internal dose assessment. An example of this is the use of the ROBY realistic rat model for the determination of red marrow *S*-values incorporating the kinetic data of the Lu-177 and Y-90-BR96 radiopharmaceuticals [81]. The rat model represented the widely used Brown Norway rats, and the correlation between absorbed dose and biological effect was evaluated. Rat computational models have been widely used in radionuclide therapy research. Thus, specific *S*-values were determined for a variety of radionuclides (Y-90, I-131, Ho-166 and Re-188) for the seven rat liver lobes [82]. A recent study on rat dosimetry simulations, presented the dependency of the rat's age in internal radiation dosimetry [83]. AFs, SAFs and *S*-values were calculated for a variety of photon and electron sources, finding that absorbed doses to organs is significantly higher in young rat models than in the adult ones. Similarly, the same authors studied the impact of obesity and emaciation of rat models on dosimetry when positron emitters are used [84]. Another recent study, investigated the dosimetry in Wistar rats of the internal exposure to neutron-activated Mn-56 dioxide powder [85].

10.3.4 Dosimetry applications in small animal models

The ICRP has provided a dose conversion factors database for a variety of reference animals and plants. On this basis, several animal models were used in ionizing radiation dosimetry. Ruedig *et al* [86] presented for first time AFs for both photon and electron sources in the range of 0.01–4.0 MeV, and 0.1–4.0 MeV, respectively, for a voxelized *Oncorhynchus mykiss*, rainbow trout. In 2008, Kinase presented a voxel based frog phantom, which was used for internal radiation dosimetry [87]. Self-AFs were evaluated in spleen, kidneys, and liver, and *S*-values were calculated for Y-90 and F-18. Based on the reference animal models by ICRP, Caffrey [88] presented voxelized models of crab and flatfish, which were used in MCNP5 for the calculation of AFs for monoenergetic electron and photon energies. In 2008, a 26.0 kg canine hybrid model was used for internal dosimetry by Padilla *et al* [89]. A dosimetric dataset was created including AFs, and SAFs, for 36 photon source organs and 30 target tissues.

10.4 Discussion

Modern nuclear medicine has managed to approach individualized dosimetry using advanced computational models and the recent developments of fast simulation techniques. A variety of high resolution animal phantoms have already been used for the assessment of accurate dosimetry parameters. In many studies the need for reference animal models, similar to reference human phantoms, is extensively reported. However, significant variations are mentioned in absorbed doses even when the anatomical differences are subtle [71, 72].

The advantages of radiation dosimetry applications using such animal computational models are numerous. First of all, realistic experimental properties are easily reproduced in the simulation 'world' with high accuracy. Researchers have all the required tools to investigate complex internal dosimetry aspects, with known ground truth provided by the predefined attenuation maps (voxel size, density, materials), the particle properties (energy type, particle interaction), and the physical processes that take place. The biokinetics of the radiopharmaceuticals, the organ's movement (cardiac and respiratory motion), and realistic environmental conditions can be defined by the users to study specific irradiation dosimetry applications. Such an approach, serves as an evaluation procedure for several dosimetry protocols, potentially translated to clinical protocols. This implicates the investigation of several parameters, without the need of irradiating and sacrificing numerous animals [90].

Nowadays, there is great interest in the scientific field of medical physics, regarding the multiscale radiation dosimetry translated in biological effects. Cellular and molecular dosimetry are of great interest, to be combined in a macroscale level (organ/tissue). In such a perspective, DNA dosimetry seems to be highly useful in preclinical studies for radiobiological research and for the investigation of the radiation impact in cancer cells and tumors, studying their physical, chemical and physico-chemical properties in animal models. Already, microdosimetry has evolved with the development of micro-therapy irradiators [91] where the use of advanced computational animal models provides additional advantages for in-depth dosimetry research.

In conclusion, the development of more and more animal models, including pets and domestic animals opens a new era in the evolution of veterinary medicine. Veterinary clinics, already use small animal imaging systems for diagnostic and therapeutic purposes. Computational models, could impact on the veterinarian's diagnosis and on the evaluation of the therapeutic protocol that will be followed.

Throughout the last three decades, it has been accepted and proved by the scientific community, that the evolution of computer science and animal computational models, has played a crucial role in multiscale, individualized dosimetry assessment.

References

[1] Xie T and Zaidi H 2016 Development of computational small animal models and their applications in preclinical imaging and therapy research *Med. Phys.* **43** 111

[2] Sarrut D, Bardies M, Boussion N, Freud N, Jan S and Letang J M *et al* 2014 A review of the use and potential of the GATE Monte Carlo simulation code for radiation therapy and dosimetry applications *Med. Phys.* **41** 064301
[3] Agostinelli S, Allison J, Amako K A and Apostolakis J *et al* 2003 GEANT4 - A simulation toolkit *Nucl. Instrum. Methods Phys. Res. section* A **506** 250–303
[4] Briesmeister J F 1986 *MCNP - A General Monte Carlo Code for Neutron and Photon Transport* (Los Alamos, NM: Los Alamos National Laboratory)
[5] Kawrakow I and Rogers D 2014 The ECSnrc code system: Monte Carlo simulations of electron and photon transport [March 31, 2018]. Available from: http://irs.inms.nrc.ca/software/egsnrc/
[6] Bernal M A, Bordage M C, Brown J M C, Davidkova M, Delage E and El Bitar Z *et al* 2015 Track structure modeling in liquid water: A review of the Geant4-DNA very low energy extension of the Geant4 Monte Carlo simulation toolkit *Phys. Med.* **31** 861–74
[7] Loevinger R, Budingeri T and Watson E E 1991 *MIRD primer for absorbed dose calculations, revised* (New York: The Society of Nuclear Medicine)
[8] Snyder W S, Ford M R and Wagner O G 1978 Estimates of specific absorbed fractions for photon sourcers uniformly distributed in various organs of a heterogeneous phantom *MIRD pamphlet no. 5* (New York: Society of Nuclear Medicine)
[9] Chetty I J, Rosu M, Kessler M L, Fraass B A, Ten Haken R K and Kong F M *et al* 2006 Reporting and analyzing statistical uncertainties in Monte Carlo-based treatment planning *Int. J. Radiat. Oncol. Biol. Phys.* **65** 1249–59
[10] Scopus. [Available from: http://scopus.com
[11] ICRU 1993 International Commission on Radiation Units and Measurements 1993 Quantities and Units in Radiation Protection Dosimetry. ICRU Report No. 51
[12] Protection ICoR 2007 Recommendations of the ICRP, ICRP Publication 103. Contract No.: 2-3
[13] Stabin M G 2016 Dosimetry of ionizing radiation in small animal imaging ed G C Kagadis, N L Forrd, D N Karnabatidis and G K Loudos *Handbook of Small Animal Imaging: Preclinical Imaging, Therapy, and Applications* (Boca Raton, London, New York: CRC Press) pp 353–64
[14] Bolch W E, Eckerman K F, Sgouros G and Thomas S R 2009 MIRD pamphlet No. 21: A generalized schema for radiopharmaceutical dosimetry--standardization of nomenclature *J. Nucl. Med.* **50** 477–84
[15] Goldman M 1982 Ionizing radiation and its risks *West J. Med.* **137** 540–7
[16] Hsieh J 2009 *Computet Tomography: Principles, Design, Artifacts and Recent Advances* 2nd edn (Bellingham, WA: SPIE)
[17] Meganck J A and Liu B 2017 Dosimetry in micro-computed tomography: A review of the measurement methods, impacts, and characterization of the quantum GX imaging system *Mol. Imaging Biol.* **19** 499–511
[18] Mahesh M 2009 *MDCT Physics: The Basics, Technology, Image Quality and Radiation Dose* (Philadelphia, PA: Lippincott Williams and Wilkins)
[19] Seeram E 2009 *Computed Tomography: Physical Principles, Clinical Applications, And Quality Control* (St. Louis, MO: Saunders, Elsevier)
[20] Tack D and Gevenois P A 2007 *Radiation Dose From Adult and Pediatric Multidetector Computed Tomography* (Berlin: Springer)
[21] Cavanaugh D, Johnson E, Price R E, Kurie J, Travis E L and Cody D D 2004 *In vivo* respiratory-gated micro-CT imaging in small-animal oncology models *Mol. Imaging* **3** 55–62

[22] Du L Y, Umoh J, Nikolov H N, Pollmann S I, Lee T Y and Holdsworth D W 2007 A quality assurance phantom for the performance evaluation of volumetric micro-CT systems *Phys. Med. Biol.* **52** 7087–108
[23] Hupfer M, Kolditz D, Nowak T, Eisa F, Brauweiler R and Kalender W A 2012 Dosimetry concepts for scanner quality assurance and tissue dose assessment in micro-CT *Med. Phys.* **39** 658–70
[24] Jacob R E, Murphy M K, Creim J A and Carson J P 2013 Detecting radiation-induced injury using rapid 3D variogram analysis of CT images of rat lungs *Acad. Radiol.* **20** 1264–71
[25] Miyahara N, Kokubo T, Hara Y, Yamada A, Koike T and Arai Y 2016 Evaluation of X-ray doses and their corresponding biological effects on experimental animals in cone-beam micro-CT scans (R-mCT2). Radiol *Phys. Technol.* **9** 60–8
[26] Vande Velde G, De Langhe E, Poelmans J, Bruyndonckx P, d'Agostino E and Verbeken E *et al* 2015 Longitudinal *in vivo* microcomputed tomography of mouse lungs: No evidence for radiotoxicity *Am. J. Physiol. Lung Cell Mol. Physiol.* **309** L271–9
[27] Willekens I, Buls N, Lahoutte T, Baeyens L, Vanhove C and Caveliers V *et al* 2010 Evaluation of the radiation dose in micro-CT with optimization of the scan protocol *Contrast Media Mol. Imaging* **5** 201–7
[28] AAPM 2008 The Measurement, Reporting, and Management of Radiation Dose in CT Report of AAPM Task Group 23 of the Diagnostic Imaging Council CT Committee. USA
[29] AAPM 1990 AAPM Report No 31: Standardized methods for measuring diagnostic X-ray exposures
[30] AAPM 1993 AAPM Report No 39: Specification and Acceptance testing of Computed Tomography scanners
[31] Jessen K, Panzer W and Shrimpton P 2000 EUR 16262: European guidelines on quality criteria for Computed Tomography. Luxembourg
[32] McCollough C H and Schueler B A 2000 Calculation of effective dose *Med. Phys.* **27** 828–37
[33] Bazalova M and Graves E E 2011 The importance of tissue segmentation for dose calculations for kilovoltage radiation therapy *Med. Phys.* **38** 3039–49
[34] Figueroa S D, Winkelmann C T, Miller H W, Volkert W A and Hoffman T J 2008 TLD assessment of mouse dosimetry during microCT imaging *Med. Phys.* **35** 3866–74
[35] Obenaus A and Smith A 2004 Radiation dose in rodent tissues during micro-CT imaging *J. X-Ray Sci. Technol.* **12** 241–9
[36] Rodt T, Luepke M, Boehm C, von Falck C, Stamm G and Borlak J *et al* 2011 Phantom and cadaver measurements of dose and dose distribution in micro-CT of the chest in mice *Acta Radiol.* **52** 75–80
[37] Boone J M, Velazquez O and Cherry S R 2004 Small-animal X-ray dose from micro-CT *Mol. Imaging* **3** 149–58
[38] Bretin F, Bahri M A, Luxen A, Phillips C, Plenevaux A and Seret A 2015 Monte Carlo simulations of the dose from imaging with GE eXplore 120 micro-CT using GATE *Med. Phys.* **42** 5711–9
[39] Jones D G and Wall B F 1985 Organ doses from medical X-ray examinations calculated using Monte Carlo techniques (Didcot: IAEA) Report NRPB-R–186
[40] Lee C, Park S, Jeon P, Jo B and Kim H 2016 Dosimetry in small-animal CT using Monte Carlo simulations *JINST* **11** T01003
[41] Taschereau R, Chow P L and Chatziioannou A F 2006 Monte carlo simulations of dose from microCT imaging procedures in a realistic mouse phantom *Med. Phys.* **33** 216–24

[42] Welch D, Turner L, Speiser M, Randers-Pehrson G and Brenner D 2017 scattered dose calculations and measurements in a life-like mouse phantom *Radiat. Res.* **187** 433–42
[43] Xu X G 2014 An exponential growth of computational phantom research in radiation protection, imaging, and radiotherapy: A review of the fifty-year history *Phys. Med. Biol.* **59** R233–302
[44] Protection ICoR 1996 Radiological protection and safety in medicine. A report of the International Commision on Radiological Protection *Ann. ICRP* **26** 1–47
[45] AAPM. AAPM position statement on radiation risks from medical imaging procedures: policy no. PP 25-B 2017 [Available from: https://aapm.org/org/policies/details.asp?id=406&type=PP
[46] Hendee W R 2013 International Organization for Medical P. Policy statement of the International Organization for Medical Physics *Radiology* **267** 326–7
[47] HPS. Health Physics Society website. Radiation risk in perspective: position statement of the Health Physics Society (PS010-1) [Available from: https://hps.org/documents/risk_ps010-2.pdf
[48] ICRP 2007 The 2007 recommendations of the International Commission on Radiological Protection. ICRP publication 103 *Ann. ICRP.* **37**
[49] UNSCEAR 2012 Report of the United Nations Scientific Committee on the Effects of Atomic Radiation (New York: United Nations)
[50] Alessio A, Manchanda V and Kinahan P 2008 Initial experience with weight-based, low-dose pediatric PET/CT protocols [abstract] *J. Nucl. Med.* **49** 85P
[51] Chawla S, Federman N and Nagata K 2008 Estimated cumulative radiation dose from PET/CT in pediatric patients with malignancies: A 5-year retrospective review [abstract] *Pediatr. Radiol.* **38** S339
[52] Gelfand M J and Lemen L C 2007 PET/CT and SPECT/CT dosimetry in children: The challenge to the pediatric imager *Semin. Nucl. Med.* **37** 391–8
[53] Lonn A 2003 Evaluation of method to minimize the effect of x-ray contrast in PET-CT attenuation correction *IEEE Nucl. Sci. Sym. Conf. Rec.* **3** 2220–1
[54] Bhargava P, He G, Samarghandi A and Delpassand E S 2012 Pictorial review of SPECT/CT imaging applications in clinical nuclear medicine *Am. J. Nucl. Med. Mol. Imaging* **2** 221–31
[55] Papadimitroulas P, Kagadis G C and Loudos G K 2016 Monte Carlo simulations in Imaging and Therapy ed G C Kagadis, N L Ford, D N Karnabatidis and G K Loudos *Handbook of Small Animal Imaging: Preclinical Imaging, Therapy, and Application* (Boca Raton, London, New York: CRC Press) pp 435–49
[56] Keenan M A, Stabin M G, Segars W P and Fernald M J 2010 RADAR realistic animal model series for dose assessment *J. Nucl. Med.* **51** 471–6
[57] Hui T E, Fisher D R, Kuhn J A, Williams L E, Nourigat C and Badger C C *et al* 1994 A mouse model for calculating cross-organ beta doses from yttrium-90-labeled immunoconjugates *Cancer.* **73** 951–7
[58] Kennel S J, Stabin M, Yoriyaz H, Brechbiel M and Mirzadeh S 1999 Treatment of lung tumor colonies with ^{90}Y targeted to blood vessels: Comparison with the alpha-particle emitter ^{213}Bi *Nucl. Med. Biol.* **26** 149–57
[59] Flynn A A, Green A J, Pedley R B, Boxer G M, Boden R and Begent R H 2001 A mouse model for calculating the absorbed beta-particle dose from (131)I-and (90)Y-labeled immunoconjugates, including a method for dealing with heterogeneity in kidney and tumor *Radiat. Res.* **156** 28–35

[60] Kolbert K S, Watson T, Matei C, Xu S, Koutcher J A and Sgouros G 2003 Murine S factors for liver, spleen, and kidney *J. Nucl. Med.* **44** 784–91
[61] Funk T, Sun M and Hasegawa B H 2004 Radiation dose estimate in small animal SPECT and PET *Med. Phys.* **31** 2680–6
[62] Hindorf C, Ljungberg M and Strand S E 2004 Evaluation of parameters influencing S values in mouse dosimetry *J. Nucl. Med.* **45** 1960–5
[63] Miller W H, Hartmann-Siantar C, Fisher D, Descalle M A, Daly T and Lehmann J *et al* 2005 Evaluation of beta-absorbed fractions in a mouse model for ^{90}Y, 188Re, ^{166}Ho, ^{149}Pm, ^{64}Cu, and ^{177}Lu radionuclides *Cancer Biother. Radiopharm.* **20** 436–49
[64] Sato H, Yamabayashi H and Nakamura T 2008 Internal dose distribution of ^{90}Y beta-ray source implanted in a small phantom simulating a mouse *Radioisotopes* **57** 385–91
[65] Stabin M G, Peterson T E, Holburn G E and Emmons M A 2006 Voxel-based mouse and rat models for internal dose calculations *J. Nucl. Med.* **47** 655–9
[66] Bitar A, Lisbona A, Thedrez P, Sai Maurel C, Le Forestier D and Barbet J *et al* 2007 A voxel-based mouse for internal dose calculations using Monte Carlo simulations (MCNP) *Phys. Med. Biol.* **52** 1013–25
[67] Chow J C, Leung M K, Lindsay P E and Jaffray D A 2010 Dosimetric variation due to the photon beam energy in the small-animal irradiation: a Monte Carlo study *Med. Phys.* **37** 5322–9
[68] Larsson E, Ljungberg M, Strand S E and Jonsson B A 2011 Monte Carlo calculations of absorbed doses in tumours using a modified MOBY mouse phantom for pre-clinical dosimetry studies *Acta Oncol.* **50** 973–80
[69] Mohammadi A and Kinase S 2011 Influence of voxel size on specific absorbed fractions and S-values in a mouse voxel phantom *Radiat. Prot. Dosimetry.* **143** 258–63
[70] Zhang X, Xie X, Cheng J, Ning Y, UYuan Y and Pan J *et al* 2012 Organ dose conversion coefficients based on a voxel mouse model and MCNP code for external photon irradiation *Radiat. Prot. Dosimetry.* **148** 9–19
[71] Xie T and Zaidi H 2013 Monte Carlo-based evaluation of S-values in mouse models for positron-emitting radionuclides *Phys. Med. Biol.* **58** 169–82
[72] Mauxion T, Barbet J, Suhard J, Pouget J P, Poirot M and Bardies M 2013 Improved realism of hybrid mouse models may not be sufficient to generate reference dosimetric data *Med. Phys.* **40** 052501
[73] Kostou T, Papadimitroulas P, Loudos G and Kagadis G C 2016 A preclinical simulated dataset of S-values and investigation of the impact of rescaled organ masses using the MOBY phantom *Phys. Med. Biol.* **61** 2333–55
[74] Retif P, Pinel S, Toussaint M, Frochot C, Chouikrat R and Bastogne T *et al* 2015 Nanoparticles for radiation therapy enhancement: The key parameters *Theranostics* **5** 1030–44
[75] Rouchota M, Loudos G, Papadimitroulas P and Kagadis G 2017 A simulation platform for optimising X-ray imaging of gold nanoparticles (#294). EMIM 2017; Cologne, Germany
[76] Konijnenberg M W, Bijster M, Krenning E P and De Jong M 2004 A stylized computational model of the rat for organ dosimetry in support of preclinical evaluations of peptide receptor radionuclide therapy with (90)Y, (111)In, or (177)Lu *J. Nucl. Med.* **45** 1260–9
[77] ICRP 2008 ICRP Publication #108: Environmental protection – The concept and use of reference animals and plants *Ann. ICRP.* **38** 25–37
[78] Peixoto P H, Vieira J W, Yoriyaz H and Lima F R 2008 Photon and electron absorbed fractions calculated from a new tomographic rat model *Phys. Med. Biol.* **53** 5343–55

[79] Wu L, Zhang G, Luo Q and Liu Q 2008 An image-based rat model for Monte Carlo organ dose calculations *Med. Phys.* **35** 3759–64
[80] Pain F, Dhenain M, Gurden H, Routier A L, Lefebvre F and Mastrippolito R *et al* 2008 A method based on Monte Carlo simulations and voxelized anatomical atlases to evaluate and correct uncertainties on radiotracer accumulation quantitation in beta microprobe studies in the rat brain *Phys. Med. Biol.* **53** 5385–404
[81] Larsson E, Ljungberg M, Martensson L, Nilsson R, Tennvall J and Strand S E *et al* 2012 Use of Monte Carlo simulations with a realistic rat phantom for examining the correlation between hematopoietic system response and red marrow absorbed dose in Brown Norway rats undergoing radionuclide therapy with (177)Lu- and (90)Y-BR96 mAbs *Med. Phys.* **39** 4434–43
[82] Xie T, Liu Q and Zaidi H 2012 Evaluation of S-values and dose distributions for (90)Y, (131) I, (166)Ho, and (188)Re in seven lobes of the rat liver *Med. Phys.* **39** 1462–72
[83] Xie T and Zaidi H 2013 Age-dependent small-animal internal radiation dosimetry *Mol. Imaging* **12** 364–75
[84] Xie T and Zaidi H 2013 Effect of emaciation and obesity on small-animal internal radiation dosimetry for positron-emitting radionuclides *Eur. J. Nucl. Med. Mol. Imaging* **40** 1748–59
[85] Stepanenko V, Rakhypbekov T, Otani K, Endo S, Satoh K and Kawano N *et al* 2017 Internal exposure to neutron-activated (56)Mn dioxide powder in Wistar rats: Part 1: Dosimetry. *Radiat. Environ. Biophys.* **56** 47–54
[86] Ruedig E, Caffrey E, Hess C and Higley K 2014 Monte Carlo derived absorbed fractions for a voxelized model of Oncorhynchus mykiss, a rainbow trout *Radiat. Environ. Biophys.* **53** 581–7
[87] Kinase S 2008 Voxel-based frog phantom for internal dose evaluation *JINST* **45** 1049–52
[88] Caffrey E 2012 Improvements in the dosimetric models of selected benthic organisms *MS Thesis* Oregon State University
[89] Padilla L, Lee C, Milner R, Shahlaee A and Bolch W E 2008 Canine anatomic phantom for preclinical dosimetry in internal emitter therapy *J. Nucl. Med.* **49** 446–52
[90] Xu X G and Eckerman K F 2010 *Handbook of Anatomical Models for Radiation Dosimetry* (Boca Raton, FL: CRC)
[91] Stojadinovic S, Low D A, Hope A J, Vicic M, Deasy J O and Cui J *et al* 2007 MicroRT-small animal conformal irradiator *Med. Phys.* **34** 4706–16

Chapter 11

Computational animal phantoms for electromagnetic dosimetry

Ilaria Liorni, Bryn Lloyd, Silvia Farcito and Niels Kuster

11.1 Introduction

In the last three decades, the use of wireless devices has become pervasive, with more than five billion devices globally today and over 80 billion forecasted for 2022 [1]. The electromagnetic (EM) fields generated by these devices couple with biological bodies of humans and animals, yet the fields induced locally in the various tissues depend on several parameters, e.g. the field strength, frequency, the reactive or radiating field, the three-dimensional (3D) field distribution (including polarization of the incident electric (E-) and magnetic (H-) fields), the shape, and the composition of the biological body. The determination of the induced fields is called 'EM dosimetry' or 'radiofrequency (RF) dosimetry'. It is well established that induced fields above established thresholds can cause unwanted nerve stimulation and/or thermal hazards [2]. However, these effects can also be used for therapeutic applications, e.g. in RF hyperthermia or in the treatment of paraplegic seizures. As insights about other potential effects below these thresholds have been sparse, and since health agencies as well as the public have clamored for conclusive data, industry (e.g. Motorola, the Mobile & Wireless Forum (MWF, formerly the Mobile Manufacturers Forum), etc) and governments (e.g. EU, USA, France, Germany, Switzerland, England, Japan, Korea, China, USA) initiated research focused programs. The majority of these programs combined epidemiology and human provocation studies, as well as *in vivo* and *in vitro* experiments [3].

In 2011, the International Agency for Research on Cancer (IARC) classified RF-fields as 'possibly carcinogenic to humans (Group 2B)' based on an association between glioma and acoustic neuroma and exposure to RF-fields reported in epidemiological studies [4]. The animal and mechanistic studies available at that time, however, provided only weak evidence for a possible relationship. Very

doi:10.1088/2053-2563/aae1b4ch11

recently, more data became available from a two year bioassay study performed as part of the National Toxicological Program (NTP), under the auspices of the National Institute of Environmental Health Sciences (NIEHS), the results of which have been classified as clear evidence for heart schwannoma and some evidence for brain glioma related to RF exposure from cell phones [5].

To date, animal studies to test health risks associated with RF exposure were performed mainly in rodents. Due to differences between humans and rodents in how the fields are coupled, detailed dosimetric evaluations are needed to be able to compare the animal exposures to that of humans and other *in vivo* studies.

Before numerical tools for performing dosimetric analysis in complex objects became available, analytical-based methods were applied to animals and humans approximated as spheres and spheroids to obtain dosimetric data [6, 7]. At the beginning of the 1990s, more refined models in the form of generic rodent models were introduced [8].

With the breakthroughs in computational power and the dawn of finite-difference time-domain (FDTD) commercial codes, several research groups started to develop computational animal models for use in RF dosimetry. The first high-resolution rat phantom was created for the numerical RF dosimetry of near-field mainly-head exposure systems operating at 900 MHz [9, 10] and later extended to 835 MHz [11, 12]. Results were carefully validated by measurements in rat cadavers with a novel 1 mm dosimetric probe. In parallel, the Brooks Air Force Base developed an anatomical model of a Sprague–Dawley rat [13, 14] which was applied for dosimetric evaluations [15, 16]. The absorbed power was estimated in the head of the rat model, which was exposed locally to a system operating at 900 MHz [15]. The results were compared to measurements made in a homogeneous phantom under the same exposure conditions as well as with the exposure in a human head model. A system for pulsed far-field whole-body EM field (EMF) exposure at 3 GHz of six animals placed in plexiglas cages with a circular antenna was also analyzed [16]. Differences in exposure as a function of animal size were first investigated with the system developed for the co-carcinogenicity study at the IRIDIUM frequency band of 1.62 GHz [17]. The Biomedical Imaging Group of the University of Southern California developed the 3D mouse atlas 'Digimouse' [18], generated by means of co-registered computed tomography (CT) and cryosection images of a 28 g nude normal male mouse, resulting in a matrix size of $380 \times 992 \times 208$ and 0.1 mm cubical voxels, with 22 structures segmented. Digimouse was used [19] to design a system for whole-body exposure at 900 MHz of unrestrained mice by taking into account the influence of the EM field.

A number of alternative mouse atlases are currently available. The 'MOBY phantom' [20] was created from the 'Visible Mouse' data from the Duke Center for *In Vivo* Microscopy. The Visible Mouse [21] is a 110 μm resolution magnetic resonance (MR) microscopy volume image of a normal 16 week-old male C57BL/6 mouse, the data of which was used to segment the model organs of the MOBY phantom. Three-dimensional non-uniform rational B-spline (NURBS) surfaces were then fit to each segmented structure to generate a mathematical phantom. Cardiac and respiratory motion were also modeled with four-dimensional (4D) NURBS in the MOBY phantom. The realistic digital rat model 'ROBY' was also developed on the basis of the NURBS technology [22].

An atlas developed at Caltech [23] contains 3D digital data based on MR microscopy of a mouse embryonic development from conception through birth. The 'Mouse Atlas Project', developed at the University of Edinburgh [24], consists of a spatial database of gene expression patterns in the developing mouse embryo.

In early 2000, the IT'IS Foundation started to develop a number of high-resolution rat and mouse models for dosimetric analysis to be used in the two-year bioassay studies designed to meet the objectives of the 'PERFORM A' project within the EUROPEAN 5th Framework program [25, 26]. The methodology used to obtain the high-resolution models, described in [27], is based on microtome slices of Sprague–Dawley rats and OF1 and B6C3F1 mice. The anatomical models reflect different genders and weights, ranging from 14–567 g in the rat and 17–46 g in the mice. Each model includes approximately 50 segmented tissues. In 2010, the models were made posable for the dosimetric analysis of the two-year cancer bioassay study of the NTP [28].

Additional models have been developed to study the absorption of RF EMF in laboratory mice exposed to radiation at 1.97 GHz and the resulting temperature increase in individual organs [29]. The image data of the new models (i.e. three-week-old male and pregnant female B6C3F1 mice at gestation day 20) were acquired in collaboration with the Animal Imaging Center (AIC) of the ETH Zurich and the University of Zurich. The AIC performed the scanning of the mice on a 9.4 T MR imaging (MRI) scanner, and the image data was segmented and processed with iSeg (ZMT Zurich Medtech AG, Zurich, Switzerland) to create a 3D surface model of the organs and tissues. A third model, that of a 23 g mouse, was created on the basis of the Digimouse atlas [18] with the addition of several missing tissues.

So far, few anatomical models have been developed for species other than rodents. The IT'IS Foundation, working jointly with the MRI+ consortium, developed a pig model based on MRI data that is used primarily as a dosimetric tool for research and development, in particular by MRI manufacturers [30]. A rabbit phantom with a resolution of 1 mm, constructed on the basis of x-ray computed tomography (CT) images taken at Kanazawa Medical University, Japan, is composed of 12 types of tissue: skin, muscle, bone, fat, brain, cerebrospinal fluid, anterior chamber, vitreous, retina/choroid/sclera, iris/ciliary body, lens, and cornea [31]. This rabbit model has been used to assess 2.45 GHz exposure to evaluate the whole-body average specific absorption rate (SAR) relationship with temperature elevation; it is also known that absorbed microwave (MW) energy causes behavioral signs of thermal stress.

In the following sections, we (i) discuss the minimal requirements for EM dosimetry of animal phantoms, including uncertainty and variation analysis, (ii) review the computational animal phantoms available in the literature in more detail, including a description of the techniques currently used to develop the numerical models, and (iii) provide an introduction to the next generation of animal models, namely, the concept of functionalized anatomical models.

11.2 Minimal requirements for EM dosimetry

The basis of the minimal requirements was outlined in [27, 32] and are summarized and expended in this section.

11.2.1 Exposure conditions

In general EM dosimetry requires the assessment of the EM field distributions inside a biological body exposed to sources that are located inside or outside the body, of which different local or integral quantities may be derived. The dosimetric requirements depend on the endpoint to be investigated, e.g. a specific organ or a specific disease. The local EM exposure quantities to be assessed can be:

- locally induced SAR;
- locally induced E-field;
- locally induced H-field;
- whole-body average SAR;
- organ-specific SAR or field;
- the local and whole-body temperature increase (ΔT).

The selection of the quantity of interest depends on the postulated interaction mechanism and some of the quantities are closely related, e.g.

$$\mathrm{SAR} = \frac{\sigma E_{\mathrm{rms}}^2}{\rho} = \lim_{t \to +0} c \frac{\partial T}{\partial t} \tag{11.1}$$

where σ is the local conductivity, ρ is the local density of the tissue, and c is the specific heat capacity of the tissue. Another requirement is that at $t = 0$, thermal equilibrium is prevalent. At very low frequencies ($f \ll 100$ kHz), the induced E-and H-fields are not correlated and at very high frequencies ($f \gg 10$ GHz), they are related to the wave impedance of the tissue.

The locally induced amplitudes of the above quantities depend on various parameters, e.g.

- animal-related:
  - outer anatomy (size/weight, age);
  - inner anatomy (size/weight, age, gender, disease state, etc);
  - tissue dielectric properties (age);
  - posture.
- exposure field-related:
  - strength of the EM field;
  - proximity to the source (reactive near-field versus far-field);
  - frequency of the EM field;
  - direction and polarization of the EM field.

It should be noted that, in general—because of these dependences—for a given exposure condition, the induced fields in the various tissues are very different and not comparable among different species (e.g. monkey, pig, cat, rat, and mouse). Even more importantly, the same applies to comparisons within the same species, which means that even within the same animal, the dosimetry is affected by, e.g. growth, aging, and changes in body weight. The exposure also instantly varies when the animal moves in the field (e.g. changing its orientation and proximity with respect to the source) or changes its exposure relevant size (e.g. changing posture, proximity to other animals, etc).

11.2.2 Animal phantoms

As localized exposure is often not possible or even desired, dosimetry must be performed for all relevant tissues. However, dosimetry can be carried out only when the following requirements for the animal phantoms are satisfied:

1. the number of detailed animal models is sufficient to well represent the entire exposure population in terms of species, gender, size/age range;
2. the resolution is much better (usually <0.1 mm) than the size of the smallest organ to be considered, with all target tissues carefully segmented; it is also important to consider the state of the organs, e.g. bladder, stomach, etc;
3. the model is posable when the animals are not immobilized during exposure;
4. the model is morphable to increase population coverage;
5. dielectric tissue parameters, average values and standard deviations, are known with sufficient precision;
6. thermal tissue parameters and perfusion models are included when thermal considerations are important.

11.2.3 Dosimetric data evaluated

For any *in vivo* study, it is good laboratory practice that the findings can be reproduced and correlated with other animal studies and compared to the human exposure. The first condition requires that the exposure system and exposure levels are described adequately. Furthermore, since the incident exposure does not correlate to human exposure or that of any other animal or other size of the same species, the dosimetry is used to provide comprehensive exposure data for all target tissues as well as sufficient information for any other tissues. This must include both the assessment of the average data, and the evaluation of the uncertainty. This latter can be estimated with Monte-Carlo methods or by splitting the total uncertainty into various sources of uncertainty—that are independent or have limited inter-dependencies—and calculations of the total uncertainty as a root-mean-squared (rms) value [32]. However, uncertainties cannot always be analyzed as independent parameters. Such combinations must be evaluated and treated accordingly, e.g. by considering the worst-case combination as a rectangular distribution. A typical list of uncertainty parameters (figure 11.1) includes all of the previously mentioned animal and exposure-specific parameters plus the uncertainty of the exposure setup, which can be split into:

- uncertainty in the numerical setup model (e.g., mechanical tolerances, material properties, source implementation, etc);
- uncertainty in the transfer sensor calibrations (e.g., linearity, pulse modulation, etc);
- numerical uncertainties.

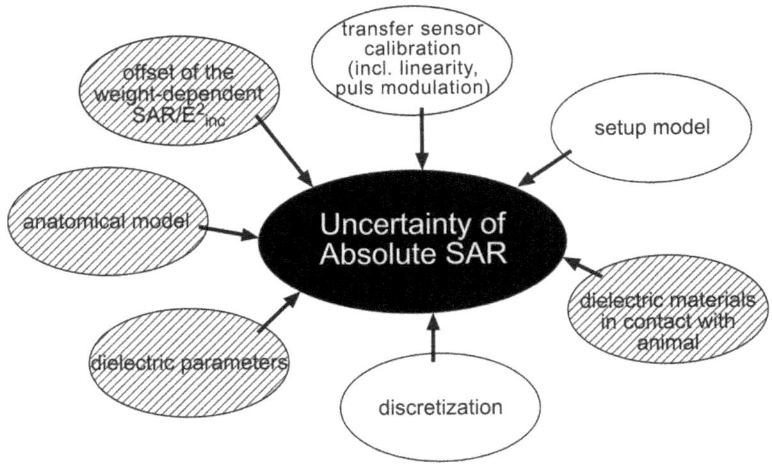

Figure 11.1. Parameters contributing to absolute SAR uncertainty (plain ellipses: apparatus, hatched ellipses: animal modeling). Note: parameters are independent of each other only when the coupling between animals is negligible [32].

In addition, the numerical evaluation needs to be experimentally validated to avoid neglecting any potential offsets. Examples of complete dosimetric evaluations have been documented [25, 33]. A substantial part of the analysis is the assessment of the uncertainty of the estimated mean value as well as the deviations from this mean value for individual animals over different averaging time periods, e.g. instantaneous (a few seconds), immediate (a week), or the entire exposure period [34].

11.2.4 Variation analysis

The exposure at any given instant (seconds) can greatly vary from the mean value over the long-term (month to life-time), e.g., due to changes in the orientation of the animal with respect to the incident field, the size and weight of the animal, the coupling between the animal and the exposure system, changes in posture, bedding, etc. Due to the potential for large variation, many additional parameters should be considered (figure 11.2) such as those related to the variations caused by the exposure setup:

- variations of the input power or incident fields due to the drifting of the measurement equipment, the scattering caused by the animal, the animal weight changes, load differences, calibration differences, non-linearities of the feedback loop, etc;
- variations of frequency (generally negligible);
- mechanical and electrical differences between the setups when several setups are employed;
- the positions occupied by the animals within the setup (applicable to setups for several restrained animals or for free moving animals).

The complexity of the analysis depends greatly on the interdependence of these parameters.

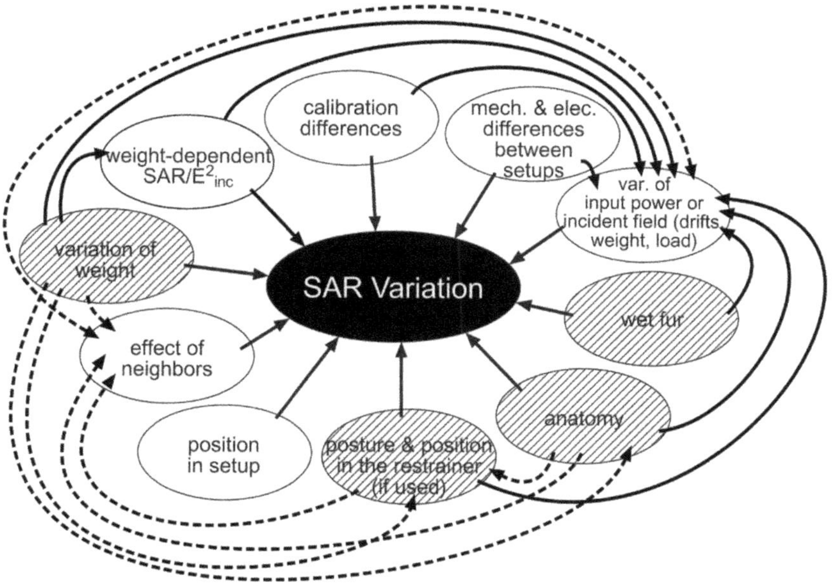

Figure 11.2. List of parameters that contribute to SAR variations (plain ellipses: apparatus; hatched ellipses: animal phantom). In the case of coupling between animals, the interaction results are more complex (dashed lines) [32].

11.3 Methods

11.3.1 Computational animal phantoms

Table 11.1 summarizes the currently existing computational phantoms of mice and rats as reported in the literature and those developed by the IT'IS Foundation (table 11.2) that belong to the Virtual Fauna or ViZoo (figures 11.3 and 11.4). In table 11.3, additional computational animal phantoms are shown. A complete list of computational animal phantoms can be found in [35].

11.3.2 Segmentation

The main steps in developing an anatomical model include:

- acquisition and pre-processing of the image data;
- segmentation of the image data to generate a label field;
- processing of the label field to remove artifacts;
- extraction of the tissue surfaces;
- processing of the surfaces;
- quality control.

The image data commonly used to generate animal models are MRI and CT data. Alternatively, microtome or cryosection slices of the animal, fixed and sectioned at constant intervals, are used.

Table 11.1. Summary of the computational mouse and rat phantoms published in peer-review journals and developed from MRI, CT, positron emission tomography (PET), and cryosection data.

Year	Reference	Affiliation	Name	Anatomical features	N° tissues	Imaging modality
MOUSE						
2004	[20]	John Hopkins University, USA	MOBY	25, 30, 35 g	14	MRI
2004	[36]	Lund University, Sweden	—	26.4 g	10	Anatomic atlas
2003	[37]	Memorial Sloan Kettering Cancer Center, USA	—	25 g	10	MRI
2006	[38]	Vanderbilt University, USA	—	27 g mouse/248 g rat	10	CT
2006	[39]	University of California, USA	—	33 g	25	CT
2007	[40]	INSERM U601, France	—	30 g	23	Cryosection
2007	[18]	University of Southern California, USA	Digimouse	28 g	22	CT, PET, cryosection
RAT						
1997	[57]	ETH Zurich, Switzerland		300 g	13	MRI
2004	[13, 14]	Brooks Air Force, USA		370 g	38	MRI
2008	[41]	University of Science and Technology, China	—	156 g	13	Cryosection
2010	[22]	John Hopkins University, USA	ROBY	200, 300, 400, 500, 600 g	14	MRI

Table 11.2. Summary of computational mouse and rat phantoms developed by the IT'IS Foundation from microtome slices and MR image data.

Year	Reference	Name	Type	Anatomical features	N° tissues	Imaging modality	
MOUSE							
2000	[25, 26]	Male PIM1	PIM1	44.7 g	49	Microtome slices	
2000	[25, 26]	Male OF1	OF1	35.5 g	50	Microtome slices	
2000	[25, 26]	Female OF1	OF1	17.3 g	48	Microtome slices	
2000	[25, 26]	Pregnant	BC63F1	28.7 g	46	Microtome slices	
2017	[18, 29]	Diggy male	Nude normal	28 g [18] Modified Diggy mouse 23 g [29]	43	CT, cryosection, PET(modification of Digimouse [18], [29])	

(*Continued*)

2017	[29]	3 week-old male mouse	BC63F1	12.3 g	50	MRI	
2017	[29]	Pregnant new model	BC63F1	38 g	52	MRI	
RAT							
2000	[25, 26]	Big Male	Sprague–Dawley	567 g	51	Microtome slices	
2000	[25, 26]	Small male	Sprague–Dawley	198 g	52	Microtome slices	
2000	[25, 26]	Female rat with tumor	Sprague–Dawley	503 g	50	Microtome slices	
2000	[25, 26]	Rat pup	Sprague–Dawley	14.3 g	49	Microtome slices	
2000	[25, 26]	Pregnant rat	Sprague–Dawley	275	52	Microtome slices	

Figure 11.3. Range of rodent computational phantoms available from the ViZoo (IT'IS Foundation).

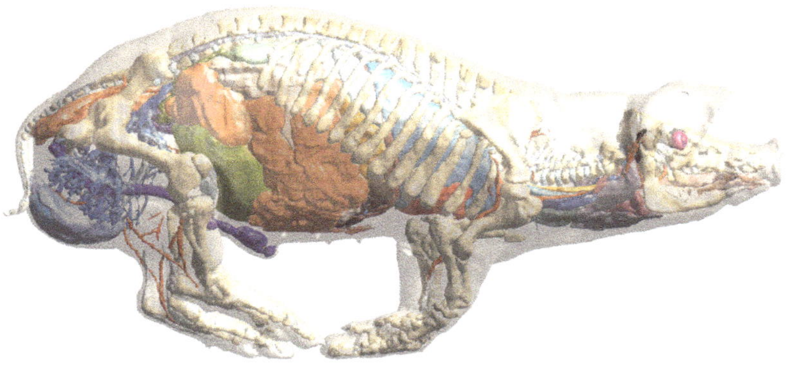

Figure 11.4. Computational pig model available from the ViZoo (IT'IS Foundation).

Depending on the resolution and quality of the acquired data, resampling can be performed to up-sample or down-sample the original datasets to reach an optimal resolution, e.g. for segmenting thin structures. During segmentation, pixels belonging to the different organs are identified on the image data and assigned to the corresponding organ/tissue.

Different image processing algorithms and segmentation tools, ranging from more manual, highly interactive to more automatic ones, can be used. For example, the ViZoo animal models have been segmented with the software iSeg [44] (ZMT Zurich MedTech AG, Zurich, Switzerland), using a range of standard techniques, such as thresholding, region growing, shape-based interpolation and manual painting. Additional specialized functions were used for cleaning and smoothing label fields, filling holes, and adding thin layers (e.g. skin).

After segmentation, the tissue boundaries are reconstructed. In this step, the tissue surfaces are extracted by means of a multi-material marching cubes algorithm [45]. Conformal surfaces offer several advantages over non-conformal ones: (i) avoidance of gaps between tissues, (ii) simplification of the creation of high-quality tetrahedral meshes, and (iii) the guarantee that structures do not overlap. The surfaces created should be smooth and the triangles not degenerated. Therefore, in the last processing step, feature- and volume-preserving smoothing is used together with a surface simplification method that should prevent the introduction of intersections and self-intersections [46]. An example of the segmentation process is shown in figure 11.5.

The whole segmentation and model generation process should be very structured to ensure that the segmentation is consistent, both within a model and among models. As part of a quality control process, detailed segmentation guidelines should be followed, together with a list of mandatory structures (organs and tissues) that must be segmented in each model according to a pre-defined tissue definition.

Table 11.3. Summary of other computational animal phantoms developed by various groups from MRI, CT and cryosection data.

Year	Animal	Reference	Affiliation	Name	Anatomical features	N° tissues	Imaging modality
2011	Male pig	[30]	IT'IS Foundation, Switzerland	—	35 000 g	103	MRI
2008	Frog	[42]	Japan Atomic Energy Agency, Japan	—	33.7 g	15	Cryosection
2008	Dog	[43]	University of Florida, US	UF	canine	26 000 g	29
CT							
2010	Rabbit	[31]	Nagoya Institute of Technology, Japan	—	2500 g	12	CT

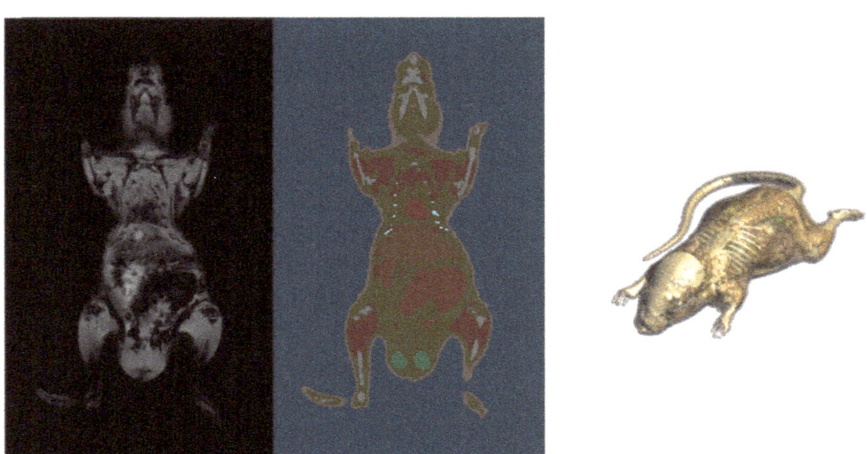

Figure 11.5. MRI scan, label-fields, and surface model (left to right) of a three-week-old male BC63F1 mouse. Models were generated according to the steps described in the segmentation section (see section 11.3.2.).

Finally, a cross-check of both the segmentation and of the generated surfaces should be performed to guarantee that the models are anatomically correct and accurate.

11.3.3 Poser

Posing functionalities allow the posture of an existing animal model to be changed to simulate more realistic exposure scenarios. Two different approaches have been developed and used until now.

In the first approach [47], the user defines a set of bones, connected via joints, as a hierarchical structure. Bones can be transformed relative to their parents, and the transformations are propagated to all children. Soft tissue deformation is computed at a vertex with the spherical blend skinning method [48] by blending bone transformations in quaternion space. The weight of each bone transformation is defined via influence regions around the bones. This technique, which allows good performance for small changes in posture, is however sensitive to the quality of the influence regions, which are difficult and time consuming to set.

The second, newer approach [49, 50] treats the body as a deformable hyper-elastic biomechanical material with rigid bones. As in the previous method, the articulation is controlled by prescribing rotations around articulated joints between the bones, and the bone hierarchy allows relative transforms to be propagated from a parent bone to all children. The new bone position is applied as a constraint to the finite element simulation. The body is discretized into soft tissue and bone tetrahedral mesh regions, which are used by the solver to compute the deformation that results from the user-defined bone position constraints. Compared to the first method, the new finite element method (FEM) based poser enforces volume preservation and results in more realistic joint-region geometries. An example of a posable rodent phantom is shown in figure 11.6.

11.4 Outlook

In this chapter, we have highlighted the importance of the use of computational animal models for EM dosimetry to reproduce laboratory experiments and investigate research questions that cannot be observed and analyzed directly *in vivo*.

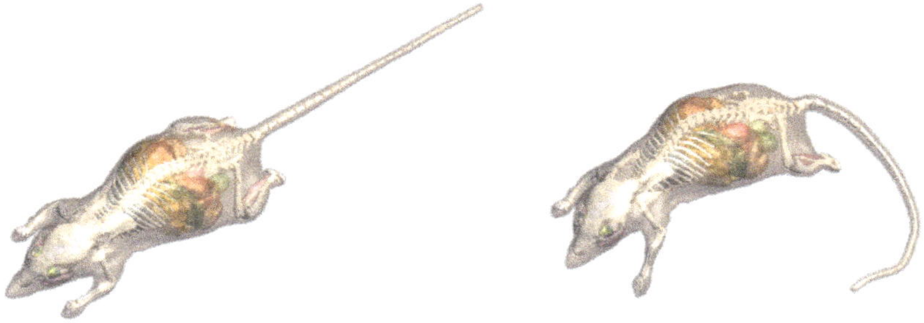

Figure 11.6. Diggy male nude normal mouse in static (left) and posed position (right). The posed model was generated based on the newer approach developed by the IT'IS Foundation as described in the posing section (see section 11.3.3).

In the future, computational animal phantoms will become more and more detailed as new high-resolution imaging techniques and registration methods become available, allowing information obtained from different modalities to be merged, or high-resolution regional models (e.g. brain atlases) to be mapped into whole-body models.

Apart from innovations in imaging methods and the creation of more detailed high-resolution phantoms, major developments can be expected in terms of functionalization of animal phantoms. Functionalization can be described as the integration of additional model layers into static, purely geometric anatomies, such as parameterized models for the beating heart and respiratory motion [20], posing and articulation [51, 52], and parameterized body-mass-index via morphing techniques [53].

Other novel forms of functionalization include the integration or assignment of physiological models to organ or tissue structures. The 'neuro-functionalization' [54, 55] of computational anatomical phantoms involves integration of neuronal electrophysiology models into the anatomical geometry. This integration is important because the non-uniform electrical conductivity of tissues surrounding neurons and/or axons leads to important non-trivial E-field distributions and gradients. Neuro-functionalized models will be indispensable for a range of new research fields and applications, for analysis of experimental findings used to build coherent understanding of the mechanisms involved in the response of biological bodies to EMF. In some cases, these mechanistic models can be used to guide the design and optimization of novel neuro-stimulation therapies, first in animals and subsequently in humans.

Two such neuro-functionalized animal phantoms are currently being developed. The first is a whole-body rat model, which will be segmented from high-resolution MRI data and will include all major tissues and organs, as well as a range of functionalized peripheral nerves, which are of interest for potential therapeutic applications (e.g. vagus nerve stimulation). The model is being created as part of the SPARC initiative, funded by the US National Institutes of Health, which has the aim to increase the understanding of how signaling from the peripheral nervous system affects the function of organs, and, consequently, how neuro-stimulation could be used to treat diseases and conditions, such as hypertension and gastrointestinal disorders. A second project, a Swiss–Korean initiative, funded by CTI in Switzerland and KIAT in South Korea, involves development of an anatomical phantom of a female macaque with neuro-functionalized peripheral nerves segmented from high-resolution color photographs of a sectioned cadaver. These efforts are directed towards tracing the trajectories of important peripheral nerves. The next level of functionalization will be achieved by mapping realistic populations of axons and tissue distributions within the nerves, to nerve trajectories in the whole body models. This information can be obtained from high-resolution image modalities, such as light-microscopy [56].

11.5 Conclusions

A large number of high-resolution animal models that have been developed for EM dosimetry are being continually refined, which will lead to increased precision of future dosimetric evaluations. The substantial dependence of the induced fields on anatomy, size and organ/tissue properties requires that detailed dosimetric assessments, including variation and uncertainty analyses, should be performed. As such, detailed dosimetric evaluations are indispensable for any animal study to meet basic scientific requirements, namely that results are comparable between studies within the same species and across different species, and are correlatable to human exposures. In general, multiple animal models of different ages, weights, and postures, as well as hundreds of simulations, are required to meet needs as outlined in this chapter. The most advanced analyses performed have been documented in [33, 34]. In the future, even more detailed evaluations will be required, for which the functionalized animal phantoms currently being developed will be indispensable.

References

[1] Cisco Visual Networking Index: Global Mobile Data Traffic Forecast Update, 2016–2021. Available at https://cisco.com/c/en/us/solutions/collateral/service-provider/visual-networking-index-vni/mobile-white-paper-c11-520862.pdf

[2] ICNIRP 1998 Guidelines for limiting exposure to time-varying electric, magnetic and electromagnetic fields (up to 300 GHz) *Health Phys.* **74** 494–522

[3] World Health Organization 2010. WHO Research Agenda for Radiofrequency Fields. Geneva, Switzerland. http://whqlibdoc.who.int/publications/2010/9789241599948_eng.pdf?ua=1

[4] IARC 2013 *Non-Ionizing radiation, Part 2: Radiofrequency electromagnetic fields. IARC* Volume 102 (Lyon, France: IARC)

[5] Guidry V NTP cell phone studies—Experts recommend elevated conclusions *Environmental Factor* April 2018 https://factor.niehs.nih.gov/2018/4/feature/feature-2-cell-phone/index.htm.

[6] Iskander M F, Massoudi H, Durney C H and Allen S J 1981 Measurements of the RF power absorption in spheroidal human and animal phantoms exposed to the near field of a dipole source *IEEE Trans. Biomed. Eng.* **28** 258–64

[7] Durney C H, Massoudi H, Iskander M F, Allen S J and Mitchell J C 1980 Radiofrequency radiation dosimetry hand-book. Technical report. 3rd edn Brooks Air Force Base: USAF School of Aerospace Medicine, Aerospace Medical Division

[8] Kuster N 1992 Immediate Report No. 4: Dosimetric Assessment for the Near-Field Exposure System. *Tech. Rep. ETH Zurich*

[9] Burkhard M, Spinelli Y and Kuster N 1997 Exposure setup to test effects of wireless communications systems on the CNS *Health Phys.* **73** 770–8

[10] Fritze K, Wiessne C, Kuster N, Sommer C, Gass P, Hermann D M, Kiessling M and Hossmann K A 1997 Effect of global system for mobile communication microwave exposure on the genomic response of the rat brain *Neuroscience.* **81** 627–39

[11] Adey W R *et al* 1999 Spontaneous and nitrosourea-induced primary tumors of the central nervous system in fischer 344 rats chronically exposed to 836 MHz modulated microwaves *Radiat. Res.* **152** 293–302

[12] Adey W R, Byus C V, Cain C D, Higgins R J, Jones R A, Kean C J, Kuster N, MacMurray A, Stagg R B and Zimmerman G 2000 Spontaneous and nitrosourea-induced primary tumors of the central nervous system in fischer 344 rats exposed to frequency-modulated microwave fields *Cancer Res.* **60** 1857–63

[13] Mason P A *et al* 1995 Database created from magnetic-resonance images of a Sprague-Dawley rat, rhesus-monkey, and pygmy goat *FASEB J.* **9** 434–40

[14] Gajsek P, Walters T J, Hurt W D, Ziriax J M, Nelson D A and Mason P A 2002 Empirical validation of SAR values predicted by FDTD modeling *Bioelectromagnetics* **23** 37–48

[15] Leveque P, Dale C, Veyret B and Wiart J 2004 Dosimetric Analysis of a 900 MHz rat head exposure system *IEEE Trans. Microw. Theory Tech.* **52** 2076–83

[16] Collin A, Perrin A, Cretallaz C, Pla S, Arnaud-Cormos D, Debouzy J C and Leveque P 2006 *In vivo* setup characterization for pulsed electromagnetic field exposure at 3 GHz *Phys. Med. Biol.* **61** 5925–41

[17] Schönborn F, Poković K and Kuster N 2004 Dosimetric analysis of the carousel setup for the exposure of rats at 1.62 GHz *Bioelectromagnetics* **25** 16–26

[18] Dogdas B, Stout D, Chatziioannou A and Leahy R M 2007 Digimouse: A 3D whole body mouse atlas from CT and cryosection data *Phys. Med. Bio* **52** 577–87

[19] Vrba J, Vrba D, Vrba J, Vožeh F, Barcal J and Vannucci L System for animal EM exposure with well defined dosimetry. *2014 General assembly and scientific symposium (URSI GASS) 2014 XXXIth URSI*

[20] Segars W P, Tsui B M W, Frey E C, Johnson G A and Berr S S 2004 Development of a 4D digital mouse phantom for molecular imaging research *Mol. Imaging Biol.* **6** 149–59

[21] Johnson G A, Cofer G P, Gewalt S L and Hedlund L W 2002 Morphologic phenotyping with magnetic resonance microscopy: the Visible Mouse *Radiology* **222** 789–93

[22] Keenan A M, Stabin M G, Segars W P and Fernald M J 2010 RADAR Realistic animal model series for dose assessment *J. Nucl. Med.* **51** 471–6

[23] Dhenain M, Ruffins S W and Jacobs R E 2001 Three-dimensional digital mouse atlas using high-resolution MRI *Dev. Biol.* **232** 458–70 (http://mouseatlas.caltech.edu/)

[24] Burger A, Baldock R and Yang Y 2002 The Edinburgh mouse atlas and gene-expression database: a spatio-temporal database for biological research *Proc. 14th Int. Conf. on Scientific and Statistical Database Management (SSDBM02)* (http://genex.hgu.mrc.ac.uk/)

[25] Tillmann T, Ernst H, Ebert S, Kuster N, Behnke W, Rittinghausen S and Dasenbrock C 2007 Carcinogenicity study of GSM and DCS wireless communication signals in B6C3F1 mice *Biolectromagnetics* **28** 173–87

[26] Smith P, Kuster N, Ebert S and Chevalier H 2007 GSM and DCS wireless communication signals: Combined chronic toxicity/carcinogenicity study in the Wistar rat *Radiat. Res.* **168** 480–92

[27] Kainz W, Nikoloski N, Oesch W, Berdinas-Torres V, Fröhlich J, Neubauer G and Kuster N 2006 Development of novel whole-body exposure setups for rats providing high efficiency, National Toxicology Program (NTP) compatibility and well-characterized exposure *Phys. Med. Biol.* **51** 5211–29

[28] Gong Y, Capstick M H, Kuehn S, Wilson P F, Ladbury J M, Koepke G, McCormick D L, Melnick R L and Niels Kuster 2007 Life-Time dosimetric assessment for mice and rats exposed in reverberation chambers for the two-year NTP cancer bioassay study on cell phone radiation *IEEE Trans. Electromagn. Compat.* **59** 1798–808

[29] Murbach M, Lloyd B, Farcito S, Schroeter A, Samaras T and Kuster N 2017 Organ-specific electrothermal dosimetry in various numerical mouse-models exposed to 1.97 GHz radio-frequency electromagnetic fields *BIOEM 2017*

[30] Van Rhoon G C, Samaras T, Yarmolenko P S, Dewhirst M W, Neufeld E and Kuster N 2013 CEM43 °C thermal dose thresholds: A potential guide for magnetic resonance radiofrequency exposure levels? *Eur. Radiol.* **23** 2215–27

[31] Hirata A, Kojima M, Kawai H, Yamashiro Y, Watanabe S, Sasaki H and Fujiwara O 2010 Acute dosimetry and estimation of threshold-inducing behavioral signs of thermal stress in rabbits at 2.45-GHz microwave exposure *IEEE Trans. Biomed. Eng.* **57** 1234–42
[32] Kuster N, Berdinas Torres V, Nikoloski N, Frauscher M and Kainz W 2006 Methodology of detailed dosimetry and treatment of uncertainty and variations for *in vivo* studies *Bioelectromagnetics* **27** 378–91
[33] Gong Y 2015 Extended methods for analysis of EM exposure for *in vitro*, *in vivo* and human studies *PhD Thesis* (Zurich: Swiss Federal Institute of Technology)
[34] Berdinas Torres V 2002 Exposure systems and dosimetry of large-scale *in vivo* studies *PhD Thesis* (Zurich: Swiss Federal Institute of Technology)
[35] Xie T and Zaidi H 2016 Development of computational small animal models and their applications in preclinical imaging and therapy research *Med. Phys.* **43** 111–31
[36] Hindorf C, Ljungberg M and Strand S E 2004 Evaluation of parameters influencing S values in mouse dosimetry *J. Nucl. Med.* **45** 1960–65
[37] Kolbert K S, Watson T, Matei C, Xu S, Koutcher A and Sgouros G 2003 Murine S factors for liver, spleen, and kidney *J. Nucl. Med.* **44** 784–91
[38] Stabin M G, Peterson T E, Holburn G E and Emmons M S 2006 Voxel-based mouse and rat models for internal dose calculations *J. Nucl. Med.* **47** 655–9
[39] Taschereau R, Chow P L and Chatziioannou A F 2006 Monte Carlo simulations of dose from microCT imaging procedures in a realistic mouse phantom *Med. Phys.* **33** 216–24
[40] Bitar A, Lisbona A, Thedrez P, Maurel C S, Le Forestier D, Barbet J and Bardies M 2007 A voxel-based mouse for internal dose calculations using Monte Carlo simulations (MCNO) *Phys. Med. Biol.* **52** 1013–25
[41] Wu L, Zhang G, Luo Q and Liu Q 2008 An image-based rat model for Monte Carlo organ dose calculations *Med. Phys.* **35** 3759–64
[42] Kinase S 2008 Voxel-based frog phantom for internal dose evaluation *J. Nucl. Sci. Technol.* **45** 1049–52
[43] Padilla L, Lee C, Milner R, Shahlaee A and Bolch W E 2008 Canine anatomic phantom for preclinical dosimetry in internal emitter therapy *J. Nucl. Med.* **49** 446–52
[44] Neufeld E 2008 High resolution hyperthermia treatment planning *PhD Thesis* (Zurich: Swiss Federal Institute of Technology)
[45] Wu Z and Sullivan J M 2003 Multiple material marching cubes algorithm *Int. J. Numer. Methods Eng.* **58** 189–207
[46] Gosselin M C *et al* 2014 Development of a new generation of high-resolution anatomical models for medical device evaluation: the Virtual Population 3.0 *Phys. Med. Biol.* **59** 5287–303
[47] Cherubini E, Chavannes N and Kuster N 2008 Realistic skeleton based deformation of high-resolution anatomical human models for electromagnetic simulation *BEMS 2008: Abstracts of the 30th Annu. Meeting of the Bioelectromagnetics Society (San Diego, CA, USA, 9–12 June)* pp 505–7 (Poster Presentation P-149 in Poster Session: Theoretical and Practical Modeling) (10 June))
[48] Ladislav K and Zara J 2005 Spherical blend skinning: a real-time deformation of articulated models *Proceed. of the 2005 Symp. on Interactive 3D graphics and games*. ACM
[49] Lloyd B, Cherubini E, Chavannes N and Kuster N 2016 Realistic physics-based posing of anatomical models for safety evaluations and computational life science in various configurations *BioEM 2016*

[50] Lloyd B, Cherubini E, Farcito S, Neufeld E, Baumgartner C and Kuster N Covering Population Variability: Morphing of Computation Anatomical Models ed *Simulation and Synthesis in Medical Imaging Proc. of the First Int. Workshop SASHIMI 2016* pp 13–22
[51] Cherubini E, Chavannes N and Kuster N 2008 Realistic skeleton based deformation of high-resolution anatomical human models for electromagnetic simulations *Proc. of the XIX General Assembly of the Int. Union of Radio Science (URSIGA 2008)*
[52] Khmelinskii A *et al* 2011 Articulated whole-body atlases for small animal image analysis: Construction and applications *Mol. Imaging Biol.* **35** 898–910
[53] Lloyd B *et al* 2016 Covering population variability: morphing of computation anatomical models *Int. Workshop on Simulation and Synthesis in Medical Imaging* (Cham: Springer)
[54] Neufeld E, Cassará A M, Montanaro H, Kuster N and Kainz W 2016 Functionalized anatomical models for EM-neuron interaction modeling *Phys. Med. Biol.* **61** 4390–401
[55] Neufeld E, Oikonomidis I, Iacono M I, Angelone L M, Kainz W and Kuster N 2016 Investigation of assumptions underlying current safety guidelines on EM-induced nerve stimulation *Phys. Med. Biol.* **61** 4466–78
[56] Gierthmuehlen M, Freiman T M, Haastert-Talini K, Mueller A, Kaminsky J, Stieglitz T and Plachta D T 2013 Computational tissue volume reconstruction of a peripheral nerve using high-resolution light-microscopy and reconstruct *PloS one* **8** e66191
[57] Burkhardt M, Spinelli Y and Kuster N 1997 Exposure setup to test effects of wireless communication systems of the CNS *Health Phys.* **73** 770–78

Chapter 12

Applications of computational animal models in imaging physics research

Rameshwar Prasad and Habib Zaidi

12.1 Introduction

Computational models involve the use of computers to simulate, study, and predict the behavior of complex systems, such as biological mechanisms within the human body, using an interdisciplinary approach involving the use of advanced mathematical, physical and computational science tools. Computational models can be defined by either mathematical equations (analytical models), matrix arrays representing voxel-based models, or by a combination of equation and voxel models (hybrid models) that mathematically describe organ contours extracted from high-resolution image voxel data. Stylized models employ regularly shaped continuous objects defined by combinations of simple surface equations, whereas voxel-based models are developed mainly from segmentation of high-resolution medical imaging tomographic data, such as computed tomography (CT) or magnetic resonance imaging (MRI) [1]. Computational models apply solid-geometry styles to represent the various anatomical features that can be assigned to various anatomical and physiological characteristics, such as biological tissue density, biochemical composition, electric properties, and, scattering and absorption coefficients for different types of electromagnetic radiation [2].

Preclinical imaging technology is playing a crucial role in biomedical research to study, image and quantify biochemical processes of animal models at the molecular level *in vivo*. The past two decades have witnessed a tremendous increase in the widespread deployment of various imaging modalities and their applications in clinical and research settings. Medical imaging modalities can be classified into two broad categories: structural (anatomic) and functional (physiologic) imaging. Ultrasound (US), bioluminescence imaging, x-ray CT, MRI provide detailed structural information, whereas single-photon emission computed tomography (SPECT), positron emission tomography (PET), fluorescence reflectance imaging,

doi:10.1088/2053-2563/aae1b4ch12

fluorescence-mediated tomography (FMT), fiber-optic microscopy, and optical frequency-domain imaging provide the biological/functional information about the subjects being imaged [3–5]. These imaging modalities are available as stand-alone systems or commonly combined in the form of dual-modality systems [6–8].

Computational models play a pivotal role in virtual simulation to mimic the internal anatomic structures and physiological characteristics of laboratory animals. Computational models are useful in imaging physics research in many ways, including but not limited to the design and optimization of new imaging systems, generation of data to develop and validate novel image acquisition and processing protocols, and assess potential improvements of new and existing image quality and quantification procedures.

Computational models enable carrying out experiments in a controlled manner, which are not feasible otherwise. The beauty of using these models is that the truth or gold standard is known in advance for comparison.

12.2 Computational animal models in imaging physics

Computational animal models, commonly known as digital phantoms, are one of the three pillars enabling accurate simulation of clinical or preclinical multimodality imaging systems. The other two pillars are the model of the image formation process and high performance computing. Computational models mirroring the anatomy or physiology of the subject under simulation are always desired since this enables simulations to be performed that are difficult or even impossible to carry out analytically or through experimental measurements [9]. There has been growing interest in the use of computational models for the simulation of laboratory animal studies which has led to the development of both analytical and voxel-based computational animal models. Recent reviews reported the availability of more than 120 computational animal models representing species of mouse, rat, crab, trout, frog, flatfish, canine, bee, deer, duck, earthworm, goat, monkey, pig, and rabbit [2].

The MOBY phantom is one of the earliest whole body computational mouse models [10]. This is a four-dimensional (4D) whole body anatomical mouse model incorporating modeling of cardiac and respiratory motion. The three-dimensional (3D) anatomy of the phantom is based on the Visible Mouse [11], a 110 micron resolution magnetic resonance microscopy volume image of a normal 16 week old male mouse data obtained from Duke University. The anatomical structures were manually segmented using the SURFdriver software [12] where internal organs were built by fitting 3D non-uniform rational B-spline (NURBS) surfaces to each segmented structure. The Digimouse atlas [13] was generated by constructing a 3D volumetric whole body mouse atlas from coregistered x-ray CT and cryosection data of a normal nude male mouse. The images were acquired post mortem from a single mouse placed on a frame with fiducials where the segmentation was performed using interactive editing tools on a wide range of anatomical structures. The atlas includes segmented structures with coregistered PET, x-ray CT and cryosection images [14]. Researchers from Huazhong University of Science and

Technology (HUST) presented a hybrid model based on cryosection images of a Sprague–Dawley (SD) rat representation [15, 16]. A computational mouse model for optical imaging simulations was presented using the boundaries of a mouse model from MRI data and discretizing the geometry by a finite element descriptor [17].

Representative computational mouse models including the stylized model of Hindorf *et al* [18], the voxel-based Digimouse model [13], and the hybrid MOBY model, [10] the first hybrid mouse model, are described in previous chapters of this volume. Advances in high-resolution PET and SPECT imaging systems [19–23], of the order of sub-millimeter spatial resolution, are enabling the development of realistic organ specific computational models, e.g. the rat brain [24].

12.3 Applications of computational animal models in imaging physics research

Realizing the crucial role of imaging in biomedical research to study, images and quantify biochemical processes at the molecular level *in vivo,* a number of dedicated scanners/cameras for laboratory animals were developed with some of them being commercially available. This wide range of imaging systems includes, but is not limited to, standalone PET, SPECT, CT, MRI, BLI, and FLI or in the form of combined or multimodality imaging systems to one another like SPECT/CT, PET/CT, PET/SPECT/CT, PET/MRI, PET/optical. Imaging physics research plays an important role in the initial design and development of imaging systems and optimization of imaging procedures, particularly when it pertains to quantification. Imaging physics research addresses the various image degrading factors faced during the image formation process and consequential parameter estimation. Computational models are good friends and indeed are a necessity for imaging physics research to provide relevant information when the ground truth is impossible to obtain. In this section, various applications of computational animal models in imaging physics are discussed.

12.3.1 Imaging systems design and performance evaluation

The design of SPECT and PET systems using the Monte Carlo method has received considerable attention and a large number of applications were reported in the literature [25, 26]. Bradshaw [25] used this tool for the design of a detector suitable for use in a SPECT cylindrically shaped scintillation camera. The detection characteristics of two scintillator materials and the optical performance of several geometric configurations were studied. The design of prototype systems that utilize solid-state detectors and low-noise electronics to achieve improved energy resolution were carried out using simulated SPECT projections of a simple myocardial perfusion phantom [27]. The results showed that a FWHM energy resolution of 3–4 keV is sufficient to render the error due to Compton scattering insignificant compared to the uncertainty due to photon statistics. New designs of imaging systems can be made and validated in any number of permutations and combinations before actual prototyping and manufacturing using simulations and computational models. Researchers from Spain reported on the virtual PET scanner where computational models were used to evaluate the performance and develop image reconstruction software long before the first real data were acquired on the prototype [28, 29]. A conceptual design of a PET camera dedicated to scanning the human brain and small animals has been presented [30]. The authors performed Monte Carlo simulations to predict the spatial resolution for a single plane PET camera with 3 mm LSO crystals. They concluded that the detector modules must be able to measure the DOI on an event-by-event basis in order to eliminate radial elongation artifacts, and that such depth information can be incorporated into the reconstruction algorithm in an artifact-free manner with a simple rebinning method [9].

Standards for performance assessment of SPECT/PET systems have been developed reporting on detailed procedures for the measurement of various parameters, such as the spatial resolution, scatter fraction, sensitivity, count losses and random coincidence measurements, and image quality characteristics [31]. In a recent study, various performance parameters were assessed using Monte Carlo simulations of the X-PET™ small-animal scanner along with cross-validation with experimental results reporting an acceptable accuracy [32]. In this work, the overall performance of the FLEX Triumph preclinical PET subsystem was assessed using the NEMA NU 4-2008 standards [31]. These standards provide a common platform for measuring the performance of various scanners using specific protocols and guidelines.

Computational models in combination with simulations can be used to design and develop novel phantoms to assess the performance parameters of imaging systems. A recent study reported on a novel cone-shaped phantom designed and fabricated for the assessment of object size-dependent scatter fraction and noise equivalent count rates for small-animal PET scanners [33]. Figure 12.1 shows a simulation model of the LabPET-8™ scanner along with the above referenced physical cone-shaped phantom.

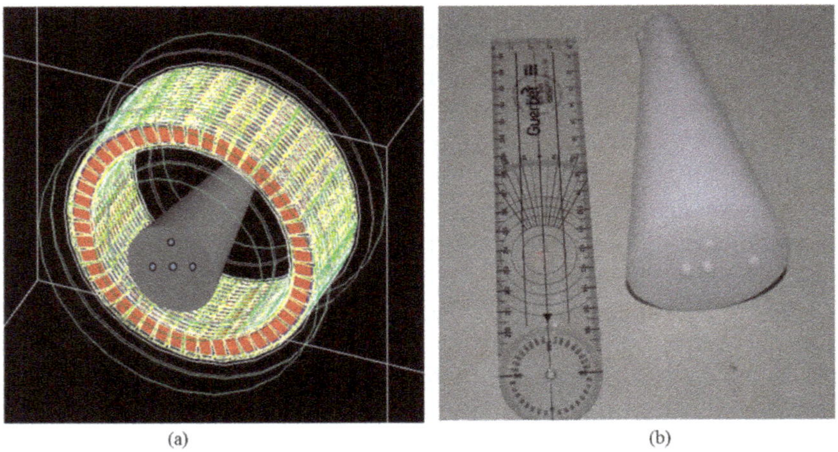

(a) (b)

Figure 12.1. Simulation model using GATE of (a) the LabPET-8™ scanner and (b) photograph of the fabricated novel cone-shaped physical phantom. Reprinted with permission from [33].

12.3.2 Modeling physical image degradation factors and their correction

Representation of the anatomical information or *in vivo* tracer distribution in organs/tissues and the underlying functional or pharmacokinetic processes taking place in the regions of interest can be estimated through accurate quantification of small-animal PET/SPECT/CT/MRI imaging data [34–40]. Extraction of quantitative parameters, such as standardized uptake value (SUV), regional glucose metabolism or parametric maps of regional cerebral blood flow, concentrations of receptors or other binding sites can be derived from PET images. The quantitative potential of PET imaging is degraded by many physical factors such as photon attenuation [41], Compton scattering [42], partial volume effect [43], and, motion [44]. These degrading factors must be corrected or compensated for to obtain accurate parameter estimates using small-animal PET imaging. These corrections can be implemented by exploiting the availability of concurrent high-resolution and contrast imaging modalities, such as x-ray CT [45] or MRI [46].

12.3.2.1 Attenuation correction

Photon attenuation in small-animal PET is not considered as major as in typical clinical imaging scenarios. However, 5 cm of soft tissue, which corresponds to the approximate transverse size of a 250 g rat, may absorb 40% of 511 keV photons [47]. This demonstrates that attenuation correction (AC) is still potentially important for accurate estimation of tracer uptake in high-resolution preclinical PET imaging [48]. Moreover, in small-animal SPECT imaging, attenuation becomes critical due to the low photon energy of commonly used radionuclides, such as technitium-99m or iodine-125 [49]. Using computational small-animal models, it can be shown that photon attenuation can reduce the measured radioactivity concentration in a volume of interest in the center of a rat-sized cylinder of water by up to 50% when imaging with iodine-125, and up to 25% when imaging with technetium-99m [50]. Information about the attenuation properties of the object being scanned can be exploited to correct for photon attenuation. Various strategies have been proposed for attenuation correction in PET [51]. Among these techniques, CT-based attenuation correction on combined PET/CT scanners is a well-established technique owing to the low statistical noise and high-quality anatomical information, small crosstalk between PET annihilation photons and low-energy x-rays, and higher throughput imaging sessions [52]. Accurate and robust conversion of CT numbers derived from low-energy polyenergetic x-ray spectra of a CT scanner to linear attenuation coefficients at 511 keV is required for correction of the corresponding PET data [53]. Several strategies have been reported in the literature, including segmentation [52], scaling [54], hybrid (segmentation/scaling) [55], bilinear or piecewise scaling [56] and dual-energy decomposition methods [57]. The bilinear conversion method assumes a linear relationship between CT numbers and linear attenuation coefficients of materials at 511 keV with one slope for soft tissue regions (air–water segment) and another slope for bone regions (water–bone segment) [58]. The impact of attenuation in estimating tumor hypoxia-related kinetic parameters for FMISO dynamic animal PET imaging was studied using a 4D dynamic digital

PET phantom of a 300 g rat with a 5 g xenografted subcutaneous human colorectal adenocarcinoma tumor [59]. Research work from Iowa University reported on the use of computational models of dog, rabbit, rat and mouse to assess the requirements in terms of attenuation correction for SPECT and PET imaging [60].

12.3.2.2 Scatter modeling and correction

Scattered events decrease image contrast by misplacing events during their assignment to a line of response, thus causing bias resulting in the overestimation of the actual activity concentration. The object scatter component is formed by annihilation photons which have interacted in the object by coherent or incoherent (primarily) processes. The object scatter distribution is object-dependent and depends on the size, depth, shape and density of the object.

The probability of Compton scattering is given by the Klein–Nishina equation, which gives the differential scattering cross section as a function of scattering angle θ:

$$\frac{d\sigma}{d\Omega} = \frac{r_e^2}{2}(1 + \cos^2\theta)\left[\frac{1}{1 + \alpha(1 - \cos\theta)}\right]^2 \left[1 + \frac{\alpha^2(1 - \cos\theta)^2}{[1 + \alpha(1 - \cos\theta)](1 + \cos^2\theta)}\right]$$

where E is the energy of the incident photon, and r_e the classical radius of the electron. This equation was widely used in the derivation of appropriate scatter models for scatter correction in PET imaging. Since all unscattered events have 511 keV before scattering, $\alpha = 1$ for the first scatter. Therefore, the differential Compton scattering cross section relative to that for unscattered annihilation photons is represented by the differential scattering probability given as:

$$\frac{d\sigma}{d\Omega} = \frac{r_e^2}{2}\left[\frac{1}{(2 - \cos\theta)^2}\right]\left[1 + \cos^2\theta + \frac{(1 - \cos\theta)^2}{(2 - \cos\theta)}\right]$$

Photons that have interacted within detector blocks by Compton scattering (53% at 511 keV for BGO) or by Rayleigh scattering (6% at 511 keV for BGO) constitute the detector scatter. These describe the detection system and are dependent upon the energy threshold discriminators. Mixed scatter is caused by photons that are scattered in both the object and the detector blocks. Mixed scatter distributions look like the projections of the object scatter convolved with the detector scatter distribution function. Various scatter-compensation techniques for PET have been proposed in the literature. They can be categorized into five main categories [61]: (i) hardware methods using septa or beam stoppers; (ii) multiple energy-window based methods; (iii) convolution/deconvolution-based methods; (iv) methods based on direct estimation of scatter distribution; and (v) methods based on statistical iterative reconstruction.

Hardware-based approaches utilize the shift in the energy discrimination window towards the high energies to reduce scatter events in the acquisition window by either using septa or beam-stopper devices made of lead [62, 63]. Multiple energy-window methods use a sufficient number of windows below and above the photopeak window to estimate the amount of scattered events [64]. Convolution/deconvolution-based techniques are based on convolution or deconvolution estimates of the distribution of scatter from the standard photopeak coincidence events. In model-based estimation of scatter distribution methods, the distribution of scattered events is commonly estimated from the emission and transmission data using either analytical or Monte Carlo modeling techniques [65]. Figure 12.2 shows the principle of model-based scatter correction. In statistical reconstruction-based methods, the scatter component is estimated during iterative reconstruction or can be generated using precomputed scatter estimators [66]. Though the magnitude of the scatter fraction (SF) has been widely studied and is well documented in clinical PET imaging, the characterization of the scatter component in small-animal PET imaging has received little attention owing to the small SF. Nevertheless, SF estimates have been reported for various small-animal PET scanners [61]. Depending on PET system characteristics and acquisition parameters, the typical range is 5%–21% for mice and 15%–30% for rats [67–74]. Although the origin of scatter for small-animal imaging has not been fully evaluated, it has been reported that the scatter mainly originates in the gantry and other hardware components rather than from the animal itself [75]. Due to the small axial field-of-view of current small-animal PET scanners relative to the length of small animals such as rats, the

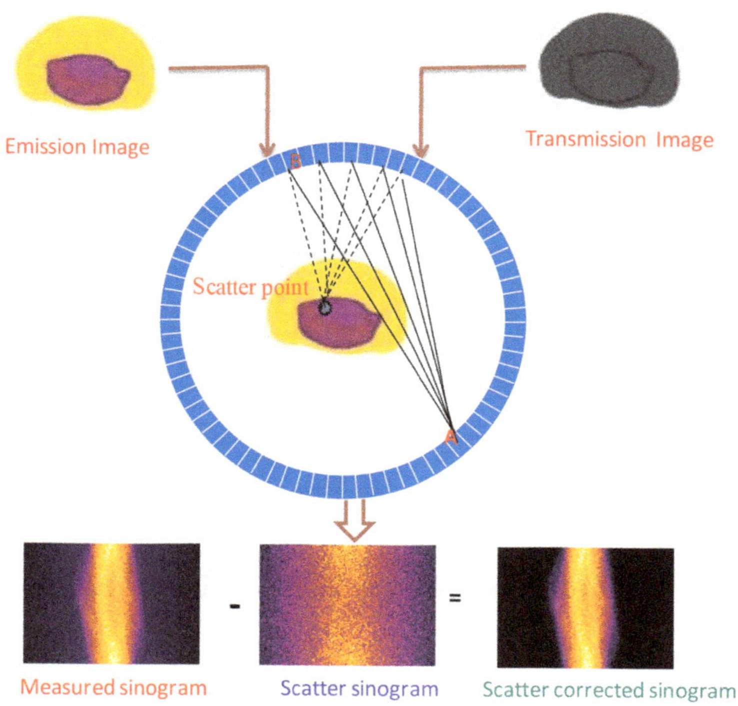

Figure 12.2. Schematic representation of the model-based scatter correction method.

whole animal cannot be scanned at one time. Hence, laboratory animals are scanned in several positions to cover the whole length. This leads to the detection of scattered events from adjacent organs while imaging the target organ. This out-of-field-of view scatter causes additional scatter and random coincidences. Monte Carlo simulation methods provide the ability to characterize the scatter originating from any source in the object being imaged, the physical components of the PET system and the remaining environment. It also provides a powerful tool to separate scattered and unscattered events as well as to control and define the phantom geometry and source position accurately. Various commonly used Monte Carlo simulation tools for PET imaging are Eidolon [76], Geant4 application for emission tomography (GATE) [77], simulation system for emission tomography (SimSET) [78], PET-SORTEO [79] and GAMOS [80]. Monte Carlo simulation techniques provide a useful tool to develop and to assess the performance and accuracy of scatter correction techniques.

Computational models of mice (e.g. MOBY) and rats (e.g. ROBY) have been used to characterize the magnitude and spatial distribution of the scatter component in small-animal PET imaging when scanning single and multiple rodents (mice and rats) simultaneously [81]. Computational modeling of simultaneous multiple objects imaging using stylized phantoms and MOBY voxelized phantoms are shown in figure 12.3. The same study has also characterized the amount of scatter caused by different sizes of small animals and the various physical parts of the scanner gantry, which is not possible to conduct experimentally. Scatter fractions of various sizes of the NEMA-mouse (diameter: 2–5.5 cm, length: 7 cm), NEMA-rat (diameter: 3–5.5 cm, length: 15 cm) and MOBY (diameter: 2.1–5.5 cm, length: 3.5–9.1 cm) phantoms were estimated using simulations with three radionuclides commonly used in small-animal SPECT studies: ^{99m}Tc (140 keV), ^{111}In (171 keV 90% and 245 keV 94%) and ^{125}I (effective energy 27.5 keV) [82].

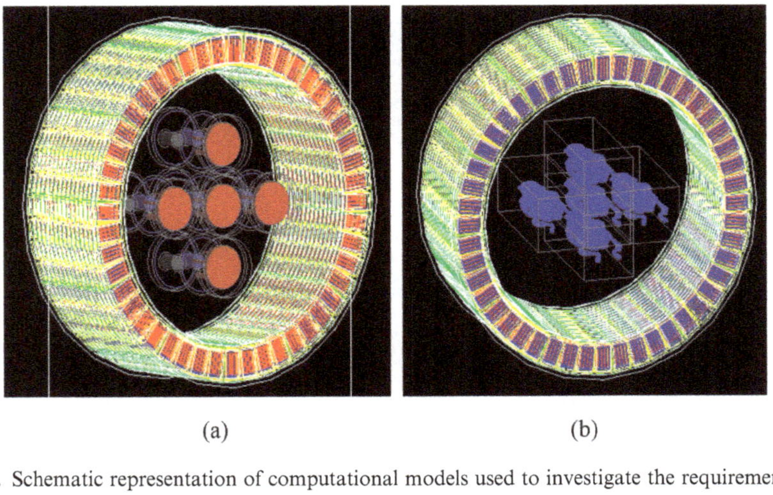

(a) (b)

Figure 12.3. Schematic representation of computational models used to investigate the requirements in terms of correction techniques for quantitative imaging of simultaneous multiple objects: (a) stylized cylindrical phantoms and (b) voxelized MOBY phantoms.

Scatter to primary ratio (SPR) was assessed using Monte Carlo calculations for the microCT subsystem of the FLEX Triumph™ preclinical PET/CT scanner and was found to be in good agreement with the experimental measurements carried out on the same scanner [83]. The impact of scatter in estimating tumor hypoxia-related kinetic parameters for FMISO dynamic animal PET imaging was studied using a 4D dynamic digital PET phantom of a 300 g rat with a 5 g xenografted subcutaneous human colorectal adenocarcinoma tumor [59].

12.3.2.3 Partial volume correction

The limited spatial resolution and signal sampling in PET imaging leads to partial volume effect (PVE) [43, 84]. PVE causes an incorrect measurement (underestimation) of radiotracer concentration especially in regions whose size is smaller than 2–3 times the spatial resolution of the PET scanner. PVE correction is especially important for dynamic quantitative studies, where aberrations in TACs due to PVE may cause bias in kinetic parameter estimation. Correction for PVE is thus required to get accurate parameter estimates [85]. Several methods were proposed for partial volume correction (PVC) in clinical PET imaging. The simpler PVC method uses recovery coefficients (RCs) [86, 87] where the correction is performed by multiplying pixel intensity values in a specific region of interest (ROI) with a size-dependent RC. RC values are determined by measuring objects with known shape containing known activity concentrations usually using physical phantom experiments. Deconvolution-based methods involve the use of the scanner's PSF and do not require aligned anatomic images [88]. Voxel-based methods use the anatomical information derived from CT or MR images to determine the spillover of activity between different regions. The PET image is segmented into different tissue types with the help of a high-resolution anatomical image (CT or MRI). Using the system point spread function (PSF) over the different tissue regions estimates the number of events originating in one tissue and detected in other tissues [89]. These methods generally work well for the brain where the assumption that tracer distribution matches regions segmented from anatomical images (white matter, gray matter and cerebrospinal fluid) is presumed. This is more problematic for whole body imaging. Recently, methods based on the multiresolution approach were developed for whole body imaging. These methods use the wavelet transform to incorporate the high-resolution information of CT or MRI into low resolution PET images [90].

Several PVC methods are validated in the context of clinical PET imaging. Some of these techniques were also exploited for small-animal PET imaging [91, 92]. Correction of the partial volume effect in mouse heart PET imaging using the fit of a 1D profile through the myocardium in gated PET images has been reported [91]. Partial volume correction in small-animal PET imaging using the continuous wavelet transform with adaptive filters was also reported [93]. PVC in the rat brain based on the geometric transfer matrix (GTM) method using anatomical regions drawn on a computational model was also investigated. In this method, partial volume effects are estimated in terms of activity retention in each region and spillover between regions is calculated by convolving each region with a measured spatially invariant point spread function [94]. A recent study evaluated partial volume effects in different sizes of rodent hearts in a small-animal PET system utilizing Monte Carlo simulation and realistic rodent computational model [95].

12.3.3 Development and evaluation of image reconstruction algorithms

Monte Carlo simulations using realistic computational models as input have been shown to be very useful for validation and comparative evaluation of image reconstruction techniques since it is possible to obtain a reference image to which reconstructed images should be compared [9]. Nowadays, it is well accepted that iterative image reconstruction methods can outperform direct methods, such as filtered-backprojection (FBP), because of their ability to incorporate an accurate physical and statistical model of the data acquisition process. Recent works have shown the high potential of using Monte Carlo simulations to derive accurate system matrices for various imaging modalities (PET, SPECT or CT) required for implementation of statistical iterative techniques [96–98]. Monte Carlo-based computation of the system matrix was used to evaluate the effect of detector response function (DRF) on reconstructed image quality [99]. Different variants of the Feldkamp (FDK) algorithm, Radon-based (RB) and shift-variant filtering (SVF) methods were evaluated using computational models in nonstandard off-centered geometries for microCT and microPET/CT [100]. A study reported on the evaluation of six reconstruction algorithms on a PET camera, namely 2D filtered FBP, 3D reprojection (3DRP), 2D maximum-likelihood expectation–maximization (MLEM), 3D ML–EM, 2D ordered subset EM (OSEM), and 3D OSEM using the computational model of the position-sensitive phototube scintillation cameras (SCs) [101]. More recent work proposed an empirical stopping criterion for the 2D MLEM iterative image reconstruction algorithm based a simulation model of a single-ring PET camera [102].

12.3.4 Quantification of small-animal PET data

Automated quantitative analysis of functional PET data is promising and provides tremendous potential enabling the practice of molecular imaging to be revolutionized since it can lower variability across centers and might enhance the consistency of image interpretation independent of user's experience. For example, the development of tracer-specific small-animal PET probabilistic atlases [103] correlated with anatomical (e.g. MRI) templates enabled automated volume of interest or voxel-based analysis of small-animal PET data with minimal user interaction [104]. One such software tool was developed by Kesner *et al* [105] to enable the assessment of the biodistribution of PET tracers using small-animal PET data. This is achieved through non-rigid coregistration of a digital mouse phantom with the animal PET image followed by automated calculation of tracer concentrations in 22 predefined VoIs representing the whole body and major organs. The development of advanced anatomical models including both stylized and more realistic voxel-based mouse [10, 13, 106] and rat [107, 108] models obtained from serial cryo-sections or dedicated high-resolution small-animal CT and MRI scanners will certainly help to support ongoing research in this area [109]. More recently, a methodology for automated atlas-guided analysis of small-animal PET data through fully automated deformable registration to an anatomical mouse model was reported [110]. Figure 12.4 shows the mouse Atlas utilised in automated quantification of functional PET data.

12.4 Summary and future directions

The popularity and widespread availability of computational models and their use in imaging physics research has grown significantly over the last two decades. This is mainly due to advances in computational power and the availability of more realistic tools for modeling boundary structures of objects. Currently, several voxel and hybrid models are available with greater realistic descriptions of anatomy and physiology of laboratory animals and humans. Computational models are aiding in the reduction of time and effort involved in the design and development of new imaging systems, corrections for physical degradation factors, and image reconstruction algorithms. This is of particular importance in biomedical research as it enables the number of laboratory animals that need to be sacrificed in various experiments to be reduced. Likewise, since computational models do not involve the handling of radioactivity, an unlimited number of experiments could be performed on a large population of subjects.

Further advances in boundary representation methods and deformable image registration techniques will translate into improving the realism of the computational models. More realistic and animal-specific models incorporating advanced respiratory and cardiac motion during imaging procedures are foreseen in the near future. Computational models are geared to play an important role in generating realistic datasets for parametric imaging studies in multimodality imaging and radiation dosimetry studies.

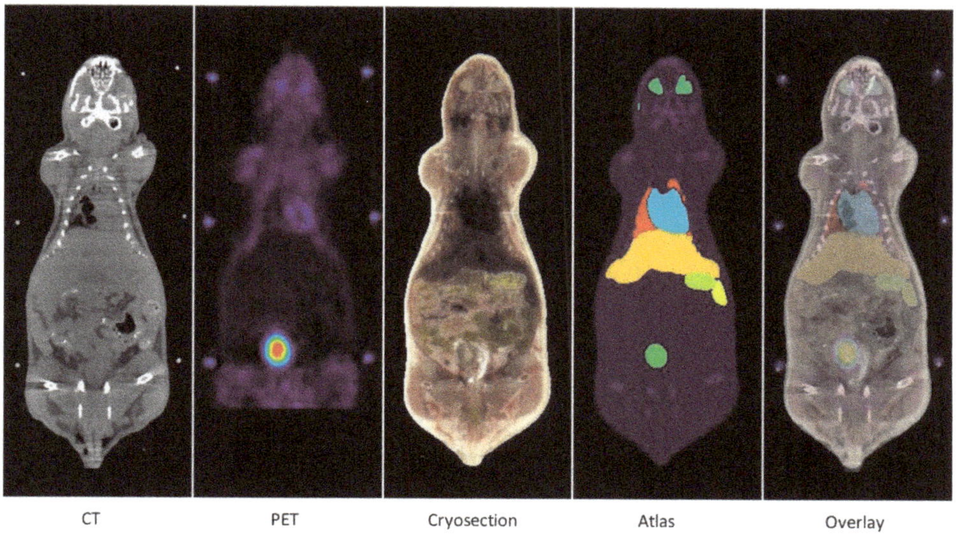

Figure 12.4. Spatially registered multimodality images for a coronal slice through the Digimouse model. From left to right: x-ray CT, PET, cryosection, segmented atlas and overlay. Reprinted with permission from [110]. © The Authors 2012.

References

[1] Zaidi H and Tsui B M W 2009 Review of computational anthropomorphic anatomical and physiological models *Proc. IEEE* **97** 1938–53

[2] Xie T and Zaidi H 2016 Development of computational small animal models and their applications in preclinical imaging and therapy research *Med. Phys.* **43** 111–31

[3] Cherry S 2004 *In vivo* molecular and genomic imaging: new challenges for imaging physics *Phys. Med. Biol.* **49** R13–48

[4] Gambhir S S 2002 Molecular imaging of cancer with positron emission tomography *Nat. Rev. Cancer* **2** 683–93

[5] Nutt R 2002 The history of positron emission tomography *Mol. Imaging Biol.* **4** 11–26

[6] Cherry S R 2006 Multimodality *in vivo* imaging systems: Twice the power or double the trouble? *Annu. Rev. Biomed. Eng.* **8** 35–62

[7] Mawlawi O and Townsend D 2009 Multimodality imaging: An update on PET/CT technology *Eur. J. Nucl. Med. Mol. Imaging* **36** (Suppl. 1) 15–29

[8] Townsend D W 2008 Multimodality imaging of structure and function *Phys. Med. Biol.* **53** R1–39

[9] Zaidi H 1999 Relevance of accurate Monte Carlo modeling in nuclear medical imaging *Med. Phys.* **26** 574–608

[10] Segars W P, Tsui B M, Frey E C, Johnson G A and Berr S S 2004 Development of a 4-D digital mouse phantom for molecular imaging research *Mol. Imaging Biol.* **6** 149–59

[11] Johnson G A, Cofer G P, Gewalt S L and Hedlund L W 2002 Morphologic phenotyping with MR microscopy: The Visible Mouse *Radiology* **222** 789–93

[12] Baiker M, Milles J, Dijkstra J, Henning T D, Weber A W and Que I *et al* 2010 Atlas-based whole-body segmentation of mice from low-contrast Micro-CT data *Med. Image Anal.* **14** 723–37

[13] Dogdas B, Stout D, Chatziioannou A F and Leahy R M 2007 Digimouse: A 3D whole body mouse atlas from CT and cryosection data *Phys. Med. Biol.* **52** 577–87

[14] Akselrod-Ballin A, Dafni H, Addadi Y, Biton I, Avni R and Brenner Y *et al* 2016 Multimodal correlative preclinical whole body imaging and segmentation *Sci. Rep.* **6** 27940

[15] Xie T, Zhang G, Li Y and Liu Q 2010 Comparison of absorbed fractions of electrons and photons using three kinds of computational phantoms of rat *Appl. Phys. Lett.* **97** 033702

[16] Zhang G, Xie T, Bosmans H and Liu Q 2009 Development of a rat computational phantom using boundary representation method for Monte Carlo simulation in radiological imaging *Proc. IEEE* **97** 2006–14

[17] Dwivedi S, Krishnan K B and Suryanarayanan S 2007 Digital mouse phantom for optical imaging *J. Biomed. Opt.* **12** 051804

[18] Hindorf C, Ljungberg M and Strand S E 2004 Evaluation of parameters influencing S values in mouse dosimetry *J. Nucl. Med.* **45** 1960–5

[19] Beekman F and van der Have F 2007 The pinhole: gateway to ultra-high-resolution three-dimensional radionuclide imaging *Eur. J. Nucl. Med. Mol. Imaging* **34** 151–61

[20] Beekman F, van der Have F and Vastenhouw B 2008 Sub-half mm rat brain SPECT with a dedicated ultra-high resolution focusing pinhole collimator J. Nucl. Med. **49** (Suppl. 1) 401P

[21] Beekman F J, van der Have F, Vastenhouw B, van der Linden A J A, van Rijk P P and Burbach J P H *et al* 2005 U-SPECT-I: A novel system for submillimeter-resolution tomography with radiolabeled molecules in mice *J. Nucl. Med.* **46** 1194–200

[22] Cherry S R 2006 The 2006 Henry N. Wagner lecture: Of mice and men (and positrons)—advances in pet imaging technology *J. Nucl. Med.* **47** 1735–45

[23] Jeavons A P, Chandler R A and Dettmar C A R 1999 A 3D HIDAC-PET camera with sub-millimetre resolution for imaging small animals *IEEE Trans. Nucl. Sci.* **46** 468–73

[24] Beekman F J and Vastenhouw B 2009 Wilt Gvd, Vervloet M, Visscher R, Booij J, *et al* 3-D rat brain phantom for high-resolution molecular imaging *Proc. IEEE* **97** 1997–2005

[25] Bradshaw J, Burnham C, Correia J, Rogers W L and Clinthorne N H 1985 Application of Monte-Carlo methods to the design of spect detector systems *IEEE Trans. Nucl. Sci.* **32** 753–7

[26] Lupton L R and Keller N A 1983 Performance study of single-slice positron emission tomography scanners by Monte Carlo techniques *IEEE Trans. Med. Imaging* **2** 154–68

[27] Heanue J A, Brown J K, Tang H R and Hasegawa B H 1996 A bound on the energy resolution required for quantitative SPECT *Med. Phys.* **23** 169–73

[28] Balcerzyk LC M, Correcher C, Gonzalez A, Vazquez C, Rubio J L, Kontaxakis G, Pozo M A and Benlloch J M 2012 Virtual PET scanner—from simulation in GATE to a final multiring Albira PET/SPECT/CT camera *Positron Emission Tomography—Current Clinical and Research Aspects* ed Chia-Hung Hsieh (Rijeka: InTech)

[29] Marcin B, George K, Mercedes D, Luis G-G, Carlos C and Antonio J G *et al* 2009 Initial performance evaluation of a high resolution Albira small animal positron emission tomography scanner with monolithic crystals and depth-of-interaction encoding from a user's perspective *Meas. Sci. Technol.* **20** 104011

[30] Miyaoka R S 1991 Dynamic high resolution positron emission imaging of rats *Biomed. Sci. Instrum.* **27** 35–42

[31] National Electrical Manufacturers Association. NEMA Standards Publication NU 4—2008. Performance Measurements of Small Animal Positron Emission Tomographs

[32] Zeraatkar N, Ay M R, Kamali-Asl A R and Zaidi H 2011 Accurate Monte Carlo modeling and performance assessment of the X-PET™ subsystem of the FLEX Triumph™ preclinical PET/CT scanner *Med. Phys.* **38** 1217–25

[33] Prasad R and Zaidi H A cone-shaped phantom for assessment of small animal PET scatter fraction and count rate performance *Mol. Imaging Bio.* **14** 561–71

[34] Bagci U, Kramer-Marek G and Mollura D J 2013 Automated computer quantification of breast cancer in small-animal models using PET-guided MR image co-segmentation. EJNMMI *Research* **3** 49

[35] Chen C-L, Wang Y, Lee J J S and Tsui B M W 2009 Toward quantitative small animal pinhole SPECT: Assessment of quantitation accuracy prior to image compensations *Mol. Imaging Bio.* **11** 195–203

[36] Comsa D C, Farrell T J and Patterson M S 2006 Quantification of bioluminescence images of point source objects using diffusion theory models *Phy. in Med. & Bio.* **51** 3733

[37] Gerdekoohi S K, Vosoughi N, Tanha K, Assadi M, Ghafarian P and Rahmim A *et al* 2017 Implementation of absolute quantification in small-animal SPECT imaging: Phantom and animal studies *J. App. Clin. Med. Phys.* **18** 215–23

[38] Grieve S M, Mazhar J, Callaghan F, Kok C Y, Tandy S and Bhindi R *et al* 2014 Automated quantification of myocardial salvage in a rat model of ischemia–reperfusion injury using 3d high-resolution magnetic resonance imaging (MRI) *J. Am. Heart Assoc.* **3**

[39] Kuntner C and Stout D 2014 Quantitative preclinical PET imaging: Opportunities and challenges *Front. Phys.* **2**

[40] Zaidi H and Sossi V 2004 Correction for image degrading factors is essential for accurate quantification of brain function using PET *Med. Phys.* **31** 423–6
[41] Zaidi H and Hasegawa B H 2003 Determination of the attenuation map in emission tomography *J. Nucl. Med.* **44** 291–315
[42] Zaidi H and Montandon M-L 2007 Scatter compensation techniques in PET *PET Clinics* **2** 219–34
[43] Rousset O, Rahmim A, Alavi A and Zaidi H 2007 Partial volume correction strategies in PET *PET Clinics* **2** 235–49
[44] Rahmim A, Rousset O and Zaidi H 2007 Strategies for motion tracking and correction in PET *PET Clinics* **2** 251–66
[45] Kalender W A 2006 X-ray computed tomography *Phys. Med. Biol.* **51** R29–43
[46] Moser E, Stadlbauer A, Windischberger C, Quick H H and Ladd M E 2008 Magnetic resonance imaging methodology *Eur. J. Nucl. Med. Mol. Imaging* **36** 30
[47] Chow P L, Rannou F R and Chatziioannou A F 2005 Attenuation correction for small animal PET tomographs *Phys. Med. Biol.* **50** 1837–50
[48] Yao R, Seidel J, Liow J-S and Green M V 2005 Attenuation correction for the NIH ATLAS small animal PET scanner *IEEE Trans. Nucl. Sci.* **52** 664–8
[49] Hwang A B and Hasegawa B H 2005 Attenuation correction for small animal SPECT imaging using x-ray CT data *Med. Phys.* **32** 2799–804
[50] Andrew B H, Benjamin L F, Grant T G and Bruce H H 2008 Assessment of the sources of error affecting the quantitative accuracy of SPECT imaging in small animals *Phys. Med. Biol.* **53** 2233
[51] Zaidi H, Montandon M-L and Alavi A 2007 Advances in attenuation correction techniques in PET *PET Clinics* **2** 191–217
[52] Kinahan P E, Hasegawa B H and Beyer T 2003 X-ray-based attenuation correction for positron emission tomography/computed tomography scanners *Semin. Nucl. Med.* **33** 166–79
[53] Prasad R, Ay M R, Ratib O and Zaidi H 2011 CT-based attenuation correction on the FLEX Triumph™ preclinical PET/CT scanner *IEEE Trans. Nucl. Sci.* **58** 66–75
[54] Beyer T, Kinahan P E, Townsend D W and Sashin D (ed) 1995 *The use of X-ray CT for attenuation correction of PET data. Proc IEEE Nuclear Science Symp. and Medical Imaging Conf. 30 Oct.-5 Nov., Norfolk, VA, USA*
[55] Kinahan P E, Townsend D W, Beyer T and Sashin D 1998 Attenuation correction for a combined 3D PET/CT scanner *Med. Phys.* **25** 2046–53
[56] Bai C, Shao L, Da Silva A J and Zhao Z 2003 A generalized model for the conversion from CT numbers to linear attenuation coefficients *IEEE Trans. Nucl. Sci.* **50** 1510–5
[57] Guy M J, Castellano-Smith I A, Flower M A, Flux G D, Ott R J and Visvikis D 1998 DETECT-dual energy transmission estimation CT-for improved attenuation correction in SPECT and PET *IEEE Trans. Nucl. Sci.* **45** 1261–7
[58] Burger C, Goerres G, Schoenes S, Buck A, Lonn A H and Von Schulthess G K 2002 PET attenuation coefficients from CT images: Experimental evaluation of the transformation of CT into PET 511-keV attenuation coefficients *Eur. J. Nucl. Med. Mol. Imaging* **29** 922–7
[59] Wang W, Chen M, Carlin S, Oehler C, Zanzonico P and Humm J 2008 Impact of attenuation and scatter correction in estimating tumor hypoxia-related kinetic parameters for FMISO dynamic animal-PET imaging *IEEE Nucl. Sci. Symp. Conf. Rec.* 5234–5239
[60] Konik A, Madsen M and Sunderland J 2009 Evaluation of attenuation correction requirement in small animal PET and SPECT imaging *J. Nucl. Med.* **50** (Suppl. 2) 1465

[61] Zaidi H and Montandon M 2007 Scatter compensation techniques in PET *PET Clinics* **2** 219–234
[62] Chuang K-S, Lin H-H, Wu J and Jan M-L 2006 Optimisation of scatter correction based on the use of partially transparent beam stoppers in PET *Nucl. Instum. Meth.* A ***569*** *175–9*
[63] Wu J, Chuang K-S, Hsu C-H, Jan M-L, Hwang I-M and Chen T-J 2005 Scatter correction for 3D PET using beam stoppers combined with dual-energy window acquisition: A feasibility study *Phys. Med. Biol.* **50** 4593–607
[64] Popescu L M, Robert M L, Samuel M and Joel S K 2006 PET energy-based scatter estimation and image reconstruction with energy-dependent corrections *Phys. Med. Biol.* **51** 2919
[65] Watson C C 2000 New, faster, image-based scatter correction for 3D PET *IEEE Trans. Nucl. Sci.* **47** 1587–94
[66] Zaidi H 2000 Comparative evaluation of scatter correction techniques in 3D positron emission tomography *Eur. J. Nucl. Med.* **27** 1813–26
[67] Bao Q, Newport D, Chen M, Stout D B and Chatziioannou A F 2009 Performance evaluation of the Inveon dedicated PET preclinical tomograph based on the NEMA NU-4 standards *J. Nucl. Med.* **50** 401–8
[68] Bergeron M, Cadorette J, Beaudoin J F, Lepage M D, Robert G and Selivanov V *et al* 2009 Performance evaluation of the LabPET APD-based digital PET scanner *IEEE Trans. Nucl. Sci.* **56** 10–6
[69] Del Guerra A, Bartoli A, Belcari N, Herbert D, Motta A and Vaiano A *et al* 2006 Performance evaluation of the fully engineered YAP-(S)PET scanner for small animal imaging *IEEE Trans. Nucl. Sci.* **53** 1078–83
[70] Kim J S, Lee J S, Im K C, Kim S J, Kim S-Y and Lee D S *et al* 2007 Performance measurement of the microPET Focus 120 scanner *J. Nucl. Med.* **48** 1527–35
[71] Prasad R, Ratib O and Zaidi H 2010 Performance evaluation of the FLEX Triumph™ X-PET scanner using the NEMA NU-04 standards *J. Nucl. Med.* **51** 1608–15
[72] Prasad R, Ratib O and Zaidi H 2011 NEMA NU-04-based performance characteristics of the LabPET-8™ small animal PET scanner *Phys. Med. Biol.* **56** 6649–64
[73] Tai Y C, Ruangma A, Rowland D, Siegel S, Newport D F and Chow P L *et al* 2005 Performance evaluation of the microPET focus: A third-generation microPET scanner dedicated to animal imaging *J. Nucl. Med.* **46** 455–63
[74] Zhang H, Bao Q, Vu N, Silverman R, Taschereau R and Berry-Pusey B *et al* 2011 Performance evaluation of PETbox: A low cost bench top preclinical PET scanner *Mol. Imaging Biol.* **13** 949–61
[75] Yang Y and Cherry S R 2006 Observations regarding scatter fraction and NEC measurements for small animal PET *IEEE Trans. Nucl. Sci.* **53** 127–32
[76] Zaidi H, Herrmann Scheurer A and Morel C 1999 An object-oriented Monte Carlo simulator for 3D positron tomographs *Comput. Meth. Prog. Biomed.* **58** 133–45
[77] Jan S, Benoit D, Becheva E, Carlier T, Cassol F and Descourt P *et al* 2011 GATE V6: A major enhancement of the GATE simulation platform enabling modelling of CT and radiotherapy *Phys. Med. Biol.* **56** 881–901
[78] SimSET. http://depts.washington.edu/simset/html/simset_main.html
[79] Reilhac A, Batan G, Michel C, Grova C, Tohka J and Collins D L *et al* 2005 PET-SORTEO: Validation and development of database of simulated PET volumes *IEEE Trans. Nucl. Sci.* **52** 1321–8

[80] Cañadas M, Arce P and Mendes P R 2010 Validation of a small-animal PET simulation using GAMOS: A GEANT4-based framework *Phys. Med. Biol.* **56** 273–88
[81] Prasad R and Zaidi H 2014 Scatter characterization and correction for simultaneous multiple small-animal PET imaging *Mol. Imaging Biol.* **16** 199–209
[82] Konik A, Madsen M T and Sunderland J J 2012 GATE simulations of small animal SPECT for determination of scatter fraction as a function of object size *IEEE Trans. Nucl. Sci.* **59** 1887–91
[83] Gutierrez D and Zaidi H 2011 Assessment of scatter for the micro-CT subsystem of the trimodality FLEX Triumph™ preclinical scanner *Med. Phys.* **38** 4154–65
[84] Soret M, Bacharach S L and Buvat I 2007 Partial-volume effect in PET tumor imaging *J. Nucl. Med.* **48** 932–45
[85] Alavi A, Werner T J, Hoilund-Carlsen P F and Zaidi H 2018 Correction for partial volume effect is a must, not a luxury, to fully exploit the potential of quantitative PET imaging in clinical oncology *Mol. Imaging Biol.* **20** 1–3
[86] Hoffman E J, Huang S C and Phelps M E 1979 Quantitation in positron emission computed tomography: 1 Effect of object size *J. Comput. Assist. Tomogr.* **3** 299–308
[87] Kessler R M, Ellis J R and Eden M 1984 Analysis of emission tomographic scan data: Limitations imposed by resolution and background *J. Comput. Assist. Tomogr.* **8** 514–22
[88] Tohka J and Reilhac A 2008 Deconvolution-based partial volume correction in Raclopride-PET and Monte Carlo comparison to MR-based method *Neuroimage* **39** 1570–84
[89] Rousset O G, Ma Y and Evans A C 1998 Correction for partial volume effects in PET: Principle and validation *J. Nucl. Med.* **39** 904–11
[90] Le Pogam A, Hatt M, Descourt P, Boussion N, Tsoumpas C and Turkheimer F E *et al* 2011 Evaluation of a 3D local multiresolution algorithm for the correction of partial volume effects in positron emission tomography *Med. Phys.* **38** 4920–3
[91] Dumouchel T and deKemp R A 2011 Analytical-based partial volume recovery in mouse heart imaging *IEEE Trans. Nucl. Sci.* **58** 110–20
[92] Julia G M, Martin S J, Andreas S, Julia T, Detlef S and Vesna S *et al* Quantification accuracy and partial volume effect in dependence of the attenuation correction of a state-of-the-art small animal PET scanner *Phys. Med. Biol.* **57** 3981
[93] Arhjoul L, Sarrhini O and Bentourkia M 2006 Partial volume correction using continuous wavelet technique in small animal PET imaging *IEEE Nucl. Sci. Symp. Rec. Oct. 29 2006-Nov. 1 2717–2721*
[94] Catana C, Drzezga A, Heiss W-D and Rosen B R 2012 PET/MRI for neurologic applications *J. Nucl. Med.* **53** 1916–25
[95] Park M-J, Fung G S K, Yamane T, Kaiser F, Fukushima K and Higuchi T 2013 Assessment of partial volume effect in small animal cardiac PET imaging using Monte Carlo simulation. *J. Nucl. Med.* **54** (Suppl. 2) 1638
[96] Leroux J-D, Selivanov V, Lecomte R and Fontaine R 2007 Accelerated iterative image reconstruction methods based on block-circulant system matrix derived from a polar image representation *IEEE Nucl. Sci. Symp. Conf. Rec. 30 Oct. - 03 Nov. 2007; Honolulu, Hawaii 2764–2771*
[97] Rafecas M, Mosler B, Dietz M, Pogl M, Stamatakis A and McElroy D P *et al* 2004 Use of a Monte Carlo-based probability matrix for 3-D iterative reconstruction of MADPET-II data *IEEE Trans. Nucl. Sci.* **51** 2597–605

[98] Rannou F R, Kohli V, Prout D L and Chatziioannou A F 2004 Investigation of OPET performance using GATE, a Geant4-based simulation software *IEEE Trans. Nucl. Sci.* **51** 2713–7

[99] Selivanov V V, Picard Y, Cadorette J, Rodrigue S and Lecomte R 2000 Detector response models for statistical iterative image reconstruction in high resolution PET *IEEE Trans. Nucl. Sci.* **47** 1168–75

[100] Valton S, Peyrin F, Delpierre P, Debarbieux F, Bonissent A and Khoury R *et al* 2007 Evaluation of tomographic reconstruction methods for small animal microCT and microPET/CT *Nucl. Instr. Meth.* A **571** 278–81

[101] Johnson C A, Seidel J, Carson R E, Gandler W R, Sofer A and Green M V *et al* 1997 Evaluation of 3D reconstruction algorithms for a small animal PET camera *IEEE Trans. Nucl. Sci.* **44** 1303–8

[102] Gaitanis A, Kontaxakis G, Spyrou G, Panayiotakis G and Tzanakos G 2010 Studying the properties of the updating coefficients in the OSEM algorithm for iterative image reconstruction in PET *Comput. Methods Programs Biomed.* **99** 219–29

[103] Casteels C, Vermaelen P, Nuyts J, Van Der Linden A, Baekelandt V and Mortelmans L *et al* 2006 Construction and evaluation of multitracer small-animal PET probabilistic atlases for voxel-based functional mapping of the rat brain *J. Nucl. Med.* **47** 1858–66

[104] Rubins D J, Melega W P, Lacan G, Way B, Plenevaux A and Luxen A *et al* 2003 Development and evaluation of an automated atlas-based image analysis method for microPET studies of the rat brain *Neuroimage* **20** 2100–18

[105] Kesner A L, Dahlbom M, Huang S C, Hsueh W A, B S P and Czernin J *et al* 2006 Semiautomated analysis of small-animal PET data *J. Nucl. Med.* **47** 1181–6

[106] Taschereau R, Chow P L and Chatziioannou A F 2006 Monte Carlo simulations of dose from microCT imaging procedures in a realistic mouse phantom *Med. Phys.* **33** 216–24

[107] Stabin M G, Peterson T E, Holburn G E and Emmons M A 2006 Voxel-based mouse and rat models for internal dose calculations *J. Nucl. Med.* **47** 655–9

[108] Wu L, Zhang G, Luo Q and Liu Q 2008 An image-based rat model for Monte Carlo organ dose calculations *Med. Phys.* **35** 3759–64

[109] Zaidi H and Xu X G 2007 Computational anthropomorphic models of the human anatomy: The path to realistic Monte Carlo modeling in medical imaging *Annu. Rev. Biomed. Eng.* **9** 471–500

[110] Gutierrez D and Zaidi H 2012 Automated analysis of small animal PET studies through deformable registration to an atlas *Eur. J. Nucl. Med. Mol. Imaging* **39** 1807–1820

Chapter 13

Applications of computational animal models in radiation therapy research

M'hamed Bentourkia, Mahdjoub Hamdi and Faiçal A A Slimani

13.1 Introduction

The importance of using small animals in biomedical research is crucial to understanding the development of various diseases and their therapies. Small animals are also used for various experiments, such as the development of biomedical instrumentation and the validation of algorithms.

Prior to investigations in small animals, scientists first work on cells in *in vitro* experiments to understand the fundamentals of cell functions, their behavior in the presence of a perturbation or the effects of a drug. Samples of cells are then easy to manipulate under controlled conditions. A further step is to assess this knowledge in naturally living cells because passive isolated cells behave differently from normal living cells within their proper physiological environment. In these conditions, the cells take advantage of their exchange of material and information, and their interaction with the immune system.

Small animal cells can then better emulate the behavior of human cells, but this is not always true due to differences in metabolism, in diets, in lifetime for cancer pathogenesis studies, and at the chromosomal and DNA levels [1–3]. Meanwhile, manipulations can be undertaken to adapt the mouse model to the proposed study such as xenotransplantation of tissues or organs, and modification of genes or immune systems [4].

To narrow the gap of the behavior of cells between small animals and humans, human diseases are implanted into humanized animals where animal genes and cells under study are replaced with their human equivalents. In certain cases, human tumors are transplanted into animals such as immunodeficient mice with the development of a functional human immune system [5–9]. Small animal models can also be used in sequential or repetitive studies and as controls for intra- or

doi:10.1088/2053-2563/aae1b4ch13

inter-subject studies. Research exclusively performed in small animals is so important that many academic laboratories and companies specialize in this area [10, 11].

Research with small animals is not limited to biology and drug development, as significant research works also focus on methodological developments linked to imaging and radiotherapy of cancer. Instead of only using phantoms to observe the provoked phenomena and to validate theoretical approaches, it is possible to perform measurements in small animals. With further studies, it is conceivable that the living animals will be replaced with computational models or digital phantoms of these animals based on atlases and on high spatial and temporal resolution imaging devices. Ultimately, the research will be safely and inexpensively conducted by computer simulations and validated against experimental measurements. The results of such research are directly transferable to experimentation in humans.

In this chapter, we will cover the usefulness and extent of performing computations using small animal models in cancer treatment with radiation therapy. Since mice are the most utilized animal models in cancer research, we restrict our discussion to the mouse model. The wide adoption of mice among other small animals is justified by their small size, in addition to the fact that they are inexpensive, easy to breed and to manipulate, reproduce rapidly, and can be genetically modified [3].

13.2 Design of digital mouse phantoms

A digital mouse can be simply designed from high resolution and high contrast computed tomography (CT) images preferably from an inert (sacrificed) mouse held in a rigid framework. CT imaging takes account of tissue density and all the parts of the mouse can be imaged. An inert mouse allows static images, without heart and lung movements which otherwise affect the spatial resolution and produce erroneous shapes of the organs in the chest and the abdomen to be acquired. For example, Dogdas *et al* co-injected a 28 g normal nude male mouse with ^{18}F-FDG and a solution of ^{18}F^{-}NaF to image most of the organs including the bones [12] (figure 13.1(a)). After 1 h, a necessary time for uptake of the radiotracers, the mouse was sacrificed and the bladder removed followed by a total freezing, then the mouse was scanned on both PET and CT scanners. The removal of the bladder was necessary to avoid measuring the accumulated radiotracers. The CT images generated a $512 \times 942 \times 512$ matrix of voxel size 0.1 mm. The same mouse was resliced by cryosection and the results of the CT, cryosection and PET produced the Digimouse mouse model. Stabin *et al* used CT scanning of a 27 g mouse to image the whole body at once providing acceptable tissue contrast and a matrix of $256 \times 256 \times 256$ with a voxel size of 0.2 mm [13] (figure 13.1(b)). Bitar *et al* froze a 30 g nude mouse, resliced it in slices of 0.1 mm then took photographs of the slices with external marker to finally form a digital mouse with voxels of $0.222 \times 0.222 \times 0.2$ mm^3 [14].

Others developed mice models from images acquired with high resolution magnetic resonance imaging (MRI). Kolbert *et al* measured a female mouse of 25 g on a 4.7 T MRI device producing 15 slices of 1.5 mm thick with 0.5 mm gap and

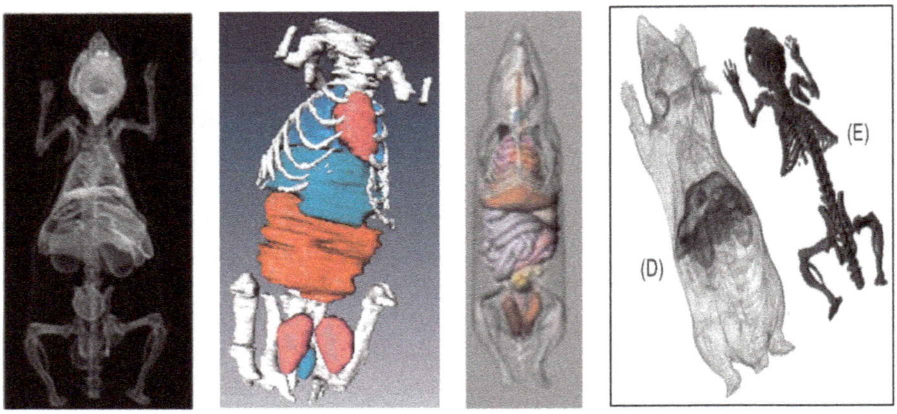

Figure 13.1. Mice models of: (a) Digimouse [12] (reproduced by permission of IOP Publishing), (b) CT-based image [13] (reproduced by permission of the Society of Nuclear Medicine and Molecular Imaging), (c) magnetic resonance microscopy (MOBY) model [25] (reproduced by permission of the Society of Nuclear Medicine and Molecular Imaging), and (d) mouse model based on Digimouse [21] (reproduced by permission from Creative Commons https://creativecommons.org/licenses/by/4.0/deed.en#).

a matrix size of 256 × 256 voxels. The voxel size was 0.01 × 0.01 × 0.2 cm^3. The 3D image was further interpolated to give 256 × 256 × 256 voxels [15]. Segars *et al* imaged a male mouse of 16 weeks of age with 3D magnetic resonance microscopy (MRM) [16]. The 3D mouse (MOBY phantom) was made of 256 × 256 × 1024 voxels with a resolution of 110 μm in all axes (figure 13.1(c)). The MOBY phantom was ingeniously designed to account for heart and lung organ motion in 12 time frames for each cardiac cycle with a voxel size of 0.1 × 0.1 × 1 mm^3, and with moving diaphragm and ribs during breathing.

Other authors proposed some enhancements or adaptations to existing models such as the MOBY mouse phantom [17–20], and the Digimouse model [21] (figure 13.1(d)). The excellent review paper by Xie and Zaidi gathered concise information from several papers which was nicely grouped in tables for comparison [22].

Based on the designed digital mouse, an atlas can be produced with homogeneous tissues assigned respective indices instead of attenuation coefficients or intensity values in the voxels [12, 13, 19, 21]. The tissues can be modified, for example, to have human tissue density and elemental composition [21, 23, 24]. The variation in voxel values, even within a homogeneous tissue, undoubtedly generates important variations of the incident particle transferred energy to tissue. This sensitivity to noise is due to the interaction of electrons set in motion by the energetic radiation, knowing that the mean free path of the ionized electrons is of the order of the mouse model voxel. Mauxion *et al* assumed the range of the radiation in imaging and radiotherapy was the same length as lengths of mice organs [19]. One has to keep in mind that a homogeneous organ such as the liver or the kidneys is not totally homogeneous, it naturally contains speckles but these are different from the noise produced during measurements with CT or MRI. Positioning of the mouse in a scanner and in a linear accelerator (LINAC) should be conformed to match the living mouse for a comparison of the results obtained in the digital and real mice.

13.3 Monte Carlo simulation platforms

X-ray beams are currently the most frequently used radiation in radiotherapy. In radiation dosimetry using small animal models, the most important parts of the calculation are the energy transfer by the x-rays in the mouse. The energy transfer can be assessed with the known probability of photon interactions with matter through Rayleigh or coherent scattering, Compton or incoherent scattering, photo-electric and pair production effects. These interaction probabilities can be deduced from the classical equations of photon interactions with matter and which can be found in several manuals, e.g. [26]. The energy transfer can be converted to absorbed dose in a voxel or in a volume of the mouse image if the radiation from the ionization electron by bremsstrahlung is taken into account. If an atomic electron absorbs the full energy of the incident photon by photoelectric effect for example at 6 MeV in liquid water, the radiation yield (fraction of kinetic energy of primary electron converted into bremsstrahlung) is around 0.0234, which can be safely neglected (https://physics.nist.gov/PhysRefData/Star/Text/ESTAR.html).

In summary, one can write a short computer program to assess the mean free path of the photon before any interaction, then decide the type of interaction and scattering angle, compute the energy deposit and follow the same photon to a geometrical or energy limit previously set, or start over with a new photon [27, 28]. Although the equations are easy to implement, some difficulties might hinder the simulation. For example one can use mono-energetic photons like in emission imaging, but this is not the case in radiation therapy where x-ray photons are produced in an x-ray tube and have to obey an energy spectrum. The uncertainties in the calculations and the computation time in the procedures are other important reasons. For these reasons and others, several platforms have been collectively developed as open source software available for academic use, usually supported by a discussion forum to debate the tools and to enhance the algorithms. These algorithms are also validated by several research groups thus increasing the degree of confidence of potential users. The major difficulty is that the user must be familiar with computer programming in order to select the appropriate functions or classes and to link them to his specific project. Below we briefly discuss some of these Monte Carlo platforms. Interested readers can find more details and examples on their web sites.

Monte Carlo simulations (MCS) are based on a sequence of random numbers for modeling stochastic phenomena deduced from known analytical formulas or probability density functions. Initial MCS considered simple investigations using mono-energetic radiation beams and uniform mediums and geometry (water, Plexiglas, spheres, ellipses, pencil beams). The increase in the number of published papers using MCS in medical physics, and the need for investigating more realistic and complex systems required a scientific collaboration to provide trusted, complete and validated MCS codes. This coincided with the wide development in computer technologies. The computers can be enhanced with graphical processing units (GPU), with a cluster of central processing units (CPU), or through a cloud of computers [29].

Several MCS packages have been written using low level computer programming languages, like C++ and Fortran, and have became available and supported by the radiation physics community [30–33]. Here we list a number of these packages: EGS4, Beam, MCNP, PENELOPE, DPM and GEANT4 (the order in which these packages are listed here is random). These codes are intended for general purpose radiation transport calculations and have a modular design, which enables the building of complex system geometries using simple shapes like boxes, spheres, … etc.

EGS4 (electron gamma shower) was written in the Fortran programming language and has a lower cutoff limit down to 1 keV for photons and 10 keV for electrons and positrons. EGS4 uses the PRESTA physics library, which combines the condensed history of the particle tracking method [34]. Using EGS4, several authors simulated heads of clinical linear accelerators by reproducing their x-ray spectra and experimental dosimetry measurements like the percentage depth doses in water phantoms. The EGSnrc package is derived from EGS4 and is provided with a graphical user interface (GUI) and improvements in the physics models, mainly with the condensed history technique, and at the photon interactions, with electron energy loss and multiple elastic scattering [35].

BEAM was originally dedicated to modeling radiotherapy treatment units by using geometrical components (slabs, stacked cones, flattening filters) [36]. It is based on EGS4 physics libraries and produces a PhaseSpace output file which could be used as an x-ray source [37]. BEAMnrc was a developed version of BEAM, which uses EGSnrc particle transport history tracking. The provided PhaseSpace files using Beam and BEAMnrc are large, making them less practical for research and clinical applications of MCS. For this reason, DOSXYZnrc was developed to calculate the absorbed dose online by compiling the PhaseSpace outputs with BEAMnrc or EGSnrc, or offline by reading the outputs from BEAMnrc or EGSnrc [38–40].

MCNP (Monte Carlo N-Particle) is a general purpose MCS code written in ANSI Fortran for neutron, photon and electron or coupled neutron/photon/electron transport [41]. It was the continuation of the works conducted in the 40s at Los Alamos National Laboratory. MCNP considers photon interactions down to 1 keV the coherent, incoherent, photoelectric, pair creation and fluorescence, annihilation and characteristic x-ray emission, and for electrons the continuous slowing down model range, ionization and bremsstrahlung. The condensed history approach is used for modeling the physics in MCNP. The cross-section of evaluated nuclear data files/libraries (ENDF/B and ENDL) consists of seven classes of nuclear cross-section data tables including neutron continuous and discrete interactions, multi-group cross-section data tables for particles coupled transport, and one class for atomic cross-section data table for photon and electron transport [41]. MCNP data output, where the history like energy deposit, elementary particle's track length, photons and electrons interaction, is defined as 'a tally' in the terminology of MCNP. MCNP provides different ways to define geometrical shapes, so instead of combing simple analytical volumes to construct complex geometrical volumes, mesh structures are used, where the whole geometrical volume surface is defined by analytical equations and elliptical tori and they are combined with logical operators, the result is called a cell. This approach provides general geometries, however, it is more difficult to simulate complex geometries like medical linear accelerators and x-ray tubes. For this reason, MCNPX (MCNP eXtension merging MCNP 4B and LAHET codes) was introduced to extensively use x-ray tubes and imaging detectors and for clinical radiotherapy and preclinical dosimetry applications [42]. MCNP/MCNPX is capable of loading and using CT or MRI datasets as voxelized geometries for external or internal dosimetry in nuclear medicine. Several GUIs were developed for dosimetry applications like SESAME and OEDIPE. These tools provide a corresponding proper MCNP input file format [43].

PENELOPE (PENetration and Energy LOss of Positrons and Electrons) is a code developed in Fortran for photon and electron transport from 1 keV to 1 GeV, especially used in accelerator beams [34, 44, 45]. As all general purpose MCS codes, a modular design is provided through an additional package called PENGEOM enabling building complex geometries based on simple homogenous volumes called 'bodies' in PENELOPE terminology [46]. Other PENELOPE derivative MCS codes, called PENEASY and PENEASYLINAC, enable automating the simulation of the Varian clinac series. These codes keep the modular design of PENELOPE and provide the user with predefined templates [47].

DPM (dose planning method) is another general purpose MCS code optimized for dose calculation using analytical or voxelized phantoms based on CT data. Like EGS4 and PENELOPE, DPM uses the mixed simulation scheme [48].

Recently, a GUI called PRIMO was developed for research and clinical applications providing predefined models of most commercial VARIAN and ELEKTA clinacs. This platform makes use of PENEASY and PENEASYLINAC for the predefined templates, PENELOPE and DPM for the tracking of photons/electrons through the linear accelerator's head and the absorbed dose in the analytical or CT data based voxelized phantoms, and also permitting delineation of tumors and organs at risk with DICOM RT-STRUCT format [49, 50].

GEANT4 (GEometry ANd Tracking) is an object oriented C++ code for tracking photons and electrons from 250 eV up to 1 TeV. This platform has its own physics models and uses the standard electromagnetic models of LIVERMORE and PENELOPE [51, 52]. Building geometry within GEANT4 is attained using imbricated arborescent volumes. The code is organized in several categories of computing classes. GEANT4 is used in many fields, such as high energy physics, space applications, radiation physics, ionizing radiation detection, medical imaging and micro-dosimetry applications [52–54]. Several MCS codes were built taking advantage of the strength of computing and performance in particle tracking of GEANT4, in addition to its availability as an open source software. As examples, GEANT4 application for tomographic emission (GATE) allows imaging and dosimetry simulation to be performed using the same physics models offered by GEANT4 and the same geometrical concepts [55, 56]; GEANT4-DNA is intended to estimate the radiation damage induced at the cellular and subcellular levels including in DNA [54].

13.4 Simulation of head of accelerators and energy spectra

13.4.1 Linear accelerator x-ray beam

To carry out a simulation in a radiation therapy environment, the design of the x-ray energy spectrum is mandatory. The x-ray beam is produced in the LINAC head. The main components of a LINAC head are the filament (cathode) emitting electrons, the high atomic number (*Z*) metal like tungsten or tungsten alloy acting as an anode to attract the accelerated electrons, and collimator and photon filtering components (figure 13.2(a)) [33, 38, 50, 57, 58]. The interaction of the accelerated electrons in the

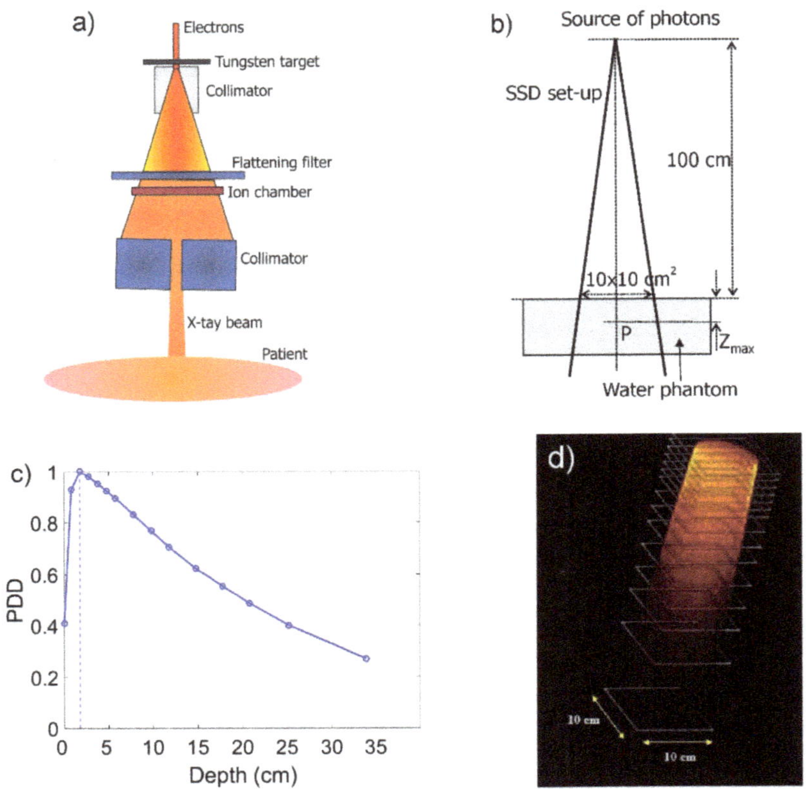

Figure 13.2. (a) Sketch of the head of a linear accelerator. (b) The x-ray source is positioned at 100 cm from the water phantom surface (source to surface distance, SSD). The dose is measured in an area of 10×10 cm^2. The percentage depth dose (PPD) is plotted as a function of depth. (c) Sketch of the PDD for 6 MeV x-ray energy as a function of depth. The maximum absorbed dose $D_{Z\max}$(Gy) = 100%PDD is at $Z_{\max}$. To calculate the absorbed dose D_Z at a given depth *Z* supposed to correspond to the depth of a tumor, from the calibration table of the accelerator, a PDD_Z for the depth *Z* is read, and D_Z(Gy) = $D_{Z\max} \times$ PDD_Z. (d) Example of MCS of the absorbed dose as a function of depth in the water phantom. The beam image was tilted to show the 10×10 cm^2 squares as a 3D view.

tungsten target generates bremsstrahlung photons of different energies. These photons form a beam which is collimated towards the patient. The low energy photons are removed with attenuating thin plates of metal like aluminum. The iron chamber allows photon fluence rate, integrated dose, and beam symmetry to be monitored. Subsequent adjustable collimators help to tailor the beam shape to the area of the tumor.

13.4.2 Calibration of the x-ray beam

In the clinic, it is difficult to adjust the filament current, the x-ray tube voltage and the collimators to set the desired absorbed dose at the desired depth in the patient. One way to determine the planned absorbed dose is to use a tank of water and to measure the dose at different depths by means of an ionization chamber (figure 13.2(b)) [59, 60]. The maximum dose as a function of depth is tagged as 100% and is generally referred to as 1 monitor unit = 1 cGy (figures 13.2(c) and (d)). From the curve of the dose versus depth, a percentage depth dose (PDD) is obtained and an absorbed dose is determined with the help of a table previously set when calibrating the LINAC. The stepping of the depth should be as fine as possible to not miss the depth at the highest dose corresponding to the peak of the PDD curve. In the clinic, the region of the highest dose in the water phantom is usually assessed at the precision of 1 mm.

13.4.3 Simulation of the x-ray beam

In order to assess the absorbed dose in the mouse, the energy spectrum needs to be known. If the energy spectrum is known from experimentation, the shape of the counts versus energy can be first fitted, and the fit is weighed based on the number of photons to be simulated for each energy of the spectrum. If the energy spectrum is not known, an MCS can be processed in the same way as the production of x-rays with an x-ray tube. Once the energy spectrum is obtained, a MCS is conducted in the water phantom in a similar manner as with the accelerator. This approach allows the number of photons and their respective energy required to provoke a given absorbed dose to be estimated. This task is more demanding in computing resources to realize because it can take days to attain the same absorbed dose as with the accelerator. Meanwhile, MCS are not expected to adequately mirror the experiments. In fact, several approaches are validated and used in MCS without being representative of all phenomena encountered in the experimentation. For example, MCS are able to retrieve the absorbed dose as measured experimentally without simulating the required photon fluence. This is accomplished by minimizing the uncertainty (variance reduction) and then by conversion of the results [59, 61]. It is good practice to assume that the absorbed dose is proportional to the photon flux, thus, simulating an adequate number of photons to achieve certain variance, and then the results are normalized to those obtained with an accelerator.

13.5 Types of absorbed doses calculated in digital mouse models

Digital mouse models are used for several types of dose calculations. These simulations can be classified into two main categories: external and internal irradiations. External and internal irradiation scenarios are commonly encountered during diagnostic or therapeutic procedures. For diagnosis, the dose received by the animal can be from external x-ray imaging like CT, or from internal radioisotope imaging, such as SPECT and PET.

In external radiation therapy, the beam can be made of electrically neutral particles, such as neutrons and photons, or of charged particles, such as electrons, protons and heavy ions. By restricting our discussion to photons, there are six main ways of administration of the beams: conventional, 3D conformal radiation therapy (3D-CRT), intensity modulated radiation therapy (IMRT), volumetric-modulated arc therapy (VMAT), and image-guided radiation therapy (IGRT) [62–66]. In all these approaches, MCS in digital mouse models can be used for various investigations. Conventional radiotherapy, or standard external radiation therapy, allows irradiation of an area of the tumor with generally a number of coplanar beams of the same or gently varying intensity between beams. 3D-CRT takes into account the volumetric shape of the tumor based on CT or MRI imaging and on the decision of the clinician to spare the organs at risk. The beams are tailored by means of collimators to match the tumor cross-section at each planned angle. 3D-CRT is adapted to the geometry of the tumor depending on the direction of the beam, and is commonly referred to as the forward treatment planning technique. Unfortunately, this approach cannot be adapted to a concave tumor, i.e. not presenting the same depth along the beam direction, or in the case where the tumor is adjacent to organs at risk. IMRT is similar to 3D-CRT in that it can adjust the intensity (photon fluence) in a single beam to different parts of the tumor. Consequently, in each beam, different parts of the tumor receive different doses. IMRT is tagged as an inverse treatment planning because, contrary to 3D-CRT where the contours are manually determined (or manually verified), the tumor is segmented based on the dose prescription to the tumor itself and to the neighboring healthy organs. In this case, algorithms implemented on the LINAC define several segments of the tumor to be treated with several appropriate doses (the contours are discussed in the following paragraph) [67, 68]. VMAT is an enhanced form of IMRT by delivering the dose to the tumor with modulated intensity and adjusted collimation to the tumor volume in a continuous rotating fashion (360°). This approach is faster and avoids higher doses to normal tissues. IGRT uses mounted CT imaging with the LINAC and takes advantage of image-guided treatment. The major benefit from IGRT is the significant positioning of the beam on the tumors as a function of time in moving organs, such as the lungs. This avenue is more interesting when it comes to adjusting the dose to tumor regions between treatment fractions, depending on the change in patient morphology, on the response of the tumor to the therapy, or for varying hypoxic regions.

Because external beams can irradiate healthy tissues in their trajectory and around the target tumor, precautions have to be taken to safely treat the tumor. The first task is to delineate the tumor as it appears on the images with a contour (gross tumor volume, GTV), then to decide the number of beams and their orientation focusing on the tumor and avoiding sensitive organs. Another region, the clinical target volume (CTV), has to be defined taking into account the potential microscopic infiltration of the tumor into the surrounding normal tissue and in the lymphatic system. A third region has also to be determined, the planning target volume (PTV) which depends on the CTV and includes uncertainties of tumor location due to organ or patient movement or errors in patient positioning. The organs at risk (OAR) have also to be delineated in order to protect the critical normal tissues from receiving significant radiation absorbed dose [68]. These 3D contours together with the beam's energy, fluence, orientation and other parameters have to be entered in the LINAC software system to form the treatment prescription. Levels of the doses are set on the different contours in this software, and finally the treatment plan is optimized.

In internal radiation therapy, the emitting source can be mechanically positioned in the tumor, or it can be transported as a molecular substance having affinity to attach to the tumor. The first case is brachytherapy where a sealed source (generally 3 mm length × 0.6 mm in diameter) is inserted within the tumor [69, 70]. Here is an example of the characteristics of two emitters used in brachytherapy, Ir-192 versus Co-60: Ir-192 decays by complex spectrum versus Co-60 which decays by 0.31 MeV β^- and by 2 γ at 1.1732 MeV and 1.3325 MeV in 99.88% of the disintegration cases, specific activity 341 GBq mg^{-1} for Ir-192 versus 42 GBq for Co-60, mean photon energy 0.355 MeV versus 1.253 MeV, half-life 74 days versus 5.27 years, source activity 370 GBq versus 74 GBq, half value layer in lead 3 mm versus 12 mm [70]. The second case concerns targeted radionuclide therapy where the radiation emitter is attached to a molecule (radiomolecule) having specific affinity to link to the targeted tumor where the radiation can be emitted [71], and in radioimmunotherapy where a radiolabeled monoclonal antibody is used to link to the target tumor [71]. Although some nanoparticles used as radiosensitizers in radiation therapy do not emit radiation, other types can be assumed as radiation emitters and/or as radiosensitizers [72, 73].

13.6 Recommendations by collaborative working groups and agencies

Several publications were collectively produced for information and educational purposes but also for providing standard guidelines based on recent published works in the field. Radiation oncologists and manufacturers of LINACs and scanners refer to these groups and agencies in their applications or developments and they assume the provided information is validated. The guidelines serve, among other benefits provided, the optimal care of patients, and the deontology and ethics of the investigators. It is desirable that clinicians and investigators use common validated and standardized tools and approaches for the calculation of the absorbed dose to allow reproducibility of results and to serve as references for investigators. Following these guidelines makes patient treatment and research universally unified.

Among these groups and agencies, which are referred to in published papers on radiation therapy and dosimetry with digital mouse models are the International Commission on Radiological Protection (ICRP), International Commission on Radiation Units and Measurements (ICRU), International Atomic Energy Agency (IAEA), Medical Internal Radiation Dose (MIRD), RAdiation Dose Assessment Resource (RADAR), and Radiation Therapy Oncology Group (RTOG).

In fact, for an MCS framework to be of interest to the scientific community, the investigator has to specifically comply with the common guidelines.

13.7 Differences between human organs and digital mouse organs in radiation therapy

Analytical calculations are easier and faster to process but they do not always reproduce the searched phenomena in human organs. The recourse to living small animals can bring more accuracy, but they also bring more complications in the experimentation regarding the differences in the physiology, the lifetime and the size. Fortunately, the development of cancer in mice is similar to that in humans, and human cancer cells can be grown in mice [1, 2, 6, 74]. Thus the effects of radiation on cancer in mice are expected to be similar to those in humans.

The experimental setup has to be adapted. In MCS, the gap between human organs and phantoms or digital animals increases since there is no cell reaction to the radiation effect in digital subjects, and there is no equivalent experimental setup. The biology and physiology are not taken into account at the moment, meanwhile, static states of human organs or living small animals can be simulated with confidence. Most of the MCS actually aim at simulating only methodologies like absorbed dose estimates as a function of other parameters of the radiation, but not, for example, the response of the tumor to the absorbed dose. Nevertheless, several assessments can be performed in digital mice models without sacrificing living mice. Hopefully, similar to the development of imaging micro-scanners, there are now dedicated small animal radiation therapy systems [75–77]. Other groups modified microCT scanners to incorporate a radiotherapy system [74, 78].

13.8 Excerpts of applications in digital mouse radiotherapy/dosimetry

Most MCS with digital mice are actually used to investigate methodologies, techniques, and calibrations. They are rarely used to investigate diseases. Because diseases and drugs are investigated in real living mice, it is worthwhile to use digital mice to quantitatively study simple phenomena which helps understanding the behavior of the diseases and the drugs. Cancer develops in mice in a similar way as in humans, and above all, human cancer cells can be transplanted into mice to study the response of cancer to therapy [1, 2, 6, 74].

To save time in MCS, some considerations can be taken depending on the goal of the simulations. If the goal of the simulation is to study the absorbed dose in a tumor and in neighboring sensitive organs with IMRT, it is not necessary to simulate the collimator and all photons interacting in the collimator since the cross-section of the tumor is already known. However, the photon energy spectra have to be accurately simulated since a single energy might produce different absorbed doses than a spectrum [79].

The tumors can be localized in particular sites in the digital mouse and the doses are prescribed. The tumor size and density are important but generally not the tumor shape. The energy of the photons is a challenging parameter since a photon from a megavoltage spectrum like in human radiotherapy has less probable interaction in the tumor's mouse, and especially the scattered photons in the tumor's mouse can have a long free path exceeding the size of the organs of the mouse. On the other hand, the use of kilovoltage photon beams can be assumed to not well represent the same damage to cells as with megavoltage photons in humans.

Some examples of recent research in radiation therapy applications in digital mice are reported in the next paragraphs. We restrict the discussion to the publications related to cancer treatment and absorbed dose.

Dose calculation with external beams to tumors of less than 1 mm in diameter and to surrounding tissues were reported in several publications, e.g. [80, 81]. Several parameters have been studied in such simulations including the absorbed dose, the beam energy and fluence, the spatial resolution and uncertainties in the tumor and organs at risk, and the comparison to parameters from measurements in human.

Simulations of absorbed dose from emitting organs within the digital mouse can fit studies in either radiopharmaceuticals from imaging devices, in diagnostic/treatment molecules (radiomolecules), or in both such as molecules emitting photons for imaging (diagnostic) and electrons for therapy (theranostic). Organ self-irradiation and organs irradiating other organs have been reported for different radiotracers and energies [20, 24, 82–85]. In such calculations, because the source organ emitting the radiation cannot have enough concentration of the radioactive elements, the dose in the organ itself and in the other organs is expected to have high uncertainty. However, the dose is generally normalized to Gy/Bq (or to Gy/Bq s for *S*-value; *S*-value or *S*-factor, expressed as $S_{t \leftarrow s}$, is the mean absorbed dose in Gy/Bq s emitted from source *s* and accumulated in target *t*). Astonishingly, the uncertainties were reported as low as 2% for *S*-values in all organs and for various radiation emitters [85].

Other authors focused their calculation on the *S*-value in various cases [14, 15, 19, 86, 87]. Calculations with radiomolecules for *S*-values could be easily substituted for calculations in brachytherapy as well as for radiosensitizing with nanoparticles in conjunction with external or internal radiation emissions [72, 88, 89] (these papers do not use digital mice models, NDMM).

Other avenues, not yet very popular, are to be used in theranostic approaches. For example, molecules or antibodies labeled with Cu-64 can be used to target specific tumors. Knowing that Cu-64 has a half-life of 12.7 h and emits positrons in 17.5% and electrons of 579 keV in 38.5% of total disintegrations, the compound can serve for imaging with PET and for treatment [90, 91] (NDMM). Labeled antibodies are more attractive since, once injected in the living body, they specifically tie to the tumor of interest and have no affinity with any other tissue. Moreover, the antibodies can directly target cell nuclei to cause damage to DNA. At the end, during the time of the radioactive emissions, only the concerned tumor is treated. This procedure is similar to brachytherapy but is less invasive and more diversified. Other MCS methods have been used to locally affect the cells. Among these methods, in 1987, Humm published a Monte Carlo approach to estimate the dose to the tumor by means of Astatine-211 emitting alpha particles with a half-life of 7.2 h [92] (NDMM). Alpha particles, being heavy and doubly charged, interact very locally in the tumor and can cause more damage either directly, by interacting with the cells, or indirectly, through ions and free radicals. A comparable approach was revisited in 2003 and the calculations were processed in living mice [93].

13.9 Conclusions

Monte Carlo simulations are easy to design and fast to run in comparison to experimental setups. However, MCS have to be first validated against experiments. Currently, a scanner or an accelerator can be exclusively designed with MCS which is an immense achievement. Unfortunately, currently MCS are not able to predict the outcome of radiation treatment in a tumor. This is because the direct and indirect effects of the radiation within the tumor cells in general and with their DNA in particular are not really known. Till now, when an amount of energy is deposited in a tumor placed in a digital mouse, the absorbed dose is translated into DNA double strand breaks assumed resulting in cell death.

Research with MCS towards DNA damage by radiation has recently been developed where Monte Carlo routines like GEANT4-DNA [54, 94, 95] have been proposed. These research works are still in the early stages and the cross sections in DNA of the interactions of incident and secondary particles in addition to the produced ions and free radicals have to be known and used in MCS. Some other effects, which should not be neglected, such as DNA repair, by standing and hypoxia have to be considered in MCS.

Regarding the computational burden of MCS, one might ask: do we really need the right number of particles necessary to exactly produce the desired absorbed dose? The answer in no because the calculations take too long and generate huge sized files if one wants to keep track of all interactions, their energies and positions. At the present time, it is suggested to simulate fewer particles till an acceptable uncertainty is achieved, then the obtained data are normalized. Another avenue is to combine short simulations with analytical distributions.

As the extreme goal, one day the digital mouse has to totally replace the real living mouse. The living mouse will only be used to generate new information to be entered in the digital mouse. On the other hand, computers, or clouds of computers, will be powerful enough to conduct complex simulations in larger bodies. But at that time, why not directly use a digital human and free the mice?

References

[1] Balmain A and Harris C C 2000 Carcinogenesis in mouse and human cells: Parallels and paradoxes *Carcinogenesis.* **21** 371–7

[2] Rangarajan A and Weinberg R A 2003 Comparative biology of mouse versus human cells: Modelling human cancer in mice *Nat. Rev. Cancer* **3** 952

[3] de Jong M and Maina T 2010 Of mice and humans: Are they the same?—Implications in cancer translational research *J. Nucl. Med.* **51** 501–4

[4] Jackson S J and Thomas G J 2017 Human tissue models in cancer research: Looking beyond the mouse *Dis. Model Mech.* **10** 939–42

[5] Clarke K, Lee F-T, Brechbiel M W, Smyth F E, Old L J and Scott A M 2000 *In vivo* biodistribution of a humanized anti-Lewis Y monoclonal antibody (hu3S193) in MCF-7 xenografted BALB/c nude mice *Cancer Res.* **60** 4804–11

[6] Wege A K, Ernst W, Eckl J, Frankenberger B, Vollmann-Zwerenz A and Männel D N *et al* 2011 Humanized tumor mice—A new model to study and manipulate the immune response in advanced cancer therapy *Int. J. Cancer* **129** 2194–206

[7] Cheon D J and Orsulic S 2011 Mouse models of cancer *Annu. Rev. Pathol.* **6** 95–119

[8] Perlman R L 2016 Mouse models of human disease: An evolutionary perspective *Evol. Med. Public Health* **2016** 170–6

[9] Zitvogel L, Pitt J M, Daillere R, Smyth M J and Kroemer G 2016 Mouse models in oncoimmunology *Nat. Rev. Cancer* **16** 759–73

[10] Verhaegen F, Granton P and Tryggestad E 2011 Small animal radiotherapy research platforms *Phys. Med. Biol.* **56** R55–83

[11] Tillner F, Thute P, Butof R, Krause M and Enghardt W 2014 Pre-clinical research in small animals using radiotherapy technology — a bidirectional translational approach *Z. Med. Phys.* **24** 335–51

[12] Dogdas B, Stout D, Chatziioannou A F and Leahy R M 2007 Digimouse: A 3D whole body mouse atlas from CT and cryosection data *Phys. Med. Biol.* **52** 577–87

[13] Stabin M G, Peterson T E, Holburn G E and Emmons M A 2006 Voxel-based mouse and rat models for internal dose calculations *J. Nucl. Med.* **47** 655–9

[14] Bitar A, Lisbona A and Bardies M 2007 S-factor calculations for mouse models using Monte-Carlo simulations *Q. J. Nucl. Med. Mol. Imaging* **51** 343–51

[15] Kolbert K S, Watson T, Matei C, Xu S, Koutcher J A and Sgouros G 2003 Murine S factors for liver, spleen, and kidney *J. Nucl. Med.* **44** 784–91

[16] Segars W P, Tsui B M W, Frey E C, Johnson G A and Berr S S 2004 Development of a 4-D digital mouse phantom for molecular imaging research *Mol. Imaging Biol.* **6** 149–59

[17] Taschereau R, Chow P L and Chatziioannou A F 2006 Monte Carlo simulations of dose from microCT imaging procedures in a realistic mouse phantom *Med. Phys.* **33** 216–24

[18] Baiker M, Milles J, Dijkstra J, Henning T D, Weber A W and Que I *et al* 2010 Atlas-based whole-body segmentation of mice from low-contrast Micro-CT data *Med. Image Anal.* **14** 723–37

[19] Mauxion T, Barbet J, Suhard J, Pouget J-P, Poirot M and Bardiès M 2013 Improved realism of hybrid mouse models may not be sufficient to generate reference dosimetric data *Med. Phys.* **40** 052501

[20] Xie T and Zaidi H 2013 Monte Carlo-based evaluation of S-values in mouse models for positron-emitting radionuclides *Phys. Med. Biol.* **58** 169–82

[21] Mendes B M, Almeida I Gd, Trindade B M, Fonseca T C F and Campos T P Rd 2017 Development of a mouse computational model for MCNPx based on Digimouse (r) images and dosimetric assays *Brazil. J. Pharm. Sci.* 53
[22] Xie T and Zaidi H 2016 Development of computational small animal models and their applications in preclinical imaging and therapy research *Med. Phys.* **43** 111–131
[23] ICRU 1991 Photon, electron, proton, and neutron interaction data for body tissues (Report 46). Bethesda
[24] Perrot Y, Degoul F, Auzeloux P, Bonnet M, Cachin F and Chezal J M *et al* 2014 Internal dosimetry through GATE simulations of preclinical radiotherapy using a melanin-targeting ligand *Phys. Med. Biol.* **59** 2183
[25] Keenan M A, Stabin M G, Segars W P and Fernald M J 2010 RADAR realistic animal model series for dose assessment *J. Nucl. Med.* **51** 471–6
[26] Evans R 1955 *The Atomic Nucleus* (New York: McGraw-Hill)
[27] Raeside D E 1976 Monte Carlo principles and applications *Phys. Med. Biol.* **21** 181
[28] Bentourkia Mh and Sarrhini O 2009 Simultaneous attenuation and scatter corrections in small animal PET imaging *Comput. Med. Imaging Graph.* **33** 477–88
[29] Ziegenhein P, Kozin I N, Kamerling C P and Oelfke U 2017 Towards real-time photon Monte Carlo dose calculation in the cloud *Phys. Med. Biol.* **62** 4375–89
[30] Andreo P 1991 Monte Carlo techniques in medical radiation physics *Phys. Med. Biol.* **36** 861
[31] Bielajew A 2001 *Fundamentals of the Monte Carlo Method for Neutral and Charged Particle Transport* (Ann Arbor, MI: The University of Michigan)
[32] Rogers D W 2006 Fifty years of Monte Carlo simulations for medical physics *Phys. Med. Biol.* **51** R287–301
[33] Chetty I J, Curran B, Cygler J E, DeMarco J J, Ezzell G and Faddegon B A *et al* 2007 Report of the AAPM Task Group No. 105: Issues associated with clinical implementation of Monte Carlo-based photon and electron external beam treatment planning *Med. Phys.* **34** 4818–53
[34] Sempau J, Acosta E, Baro J, Fernández-Varea J M and Salvat F 1997 An algorithm for Monte Carlo simulation of coupled electron-photon transport *Nucl. Instrum. Meth. Phys. Res. B* **132** 377–90
[35] Kawrakow I 2000 Accurate condensed history Monte Carlo simulation of electron transport. I. EGSnrc, the new EGS4 version *Med. Phys.* **27** 485–98
[36] Rogers D W, Faddegon B A, Ding G X, Ma C M, We J and Mackie T R 1995 BEAM: A Monte Carlo code to simulate radiotherapy treatment units *Med. Phys.* **22** 503–24
[37] Sheikh-Bagheri D and Rogers D W 2002 Monte Carlo calculation of nine megavoltage photon beam spectra using the BEAM code *Med. Phys.* **29** 391–402
[38] Kawrakow I and Walters B R 2006 Efficient photon beam dose calculations using DOSXYZnrc with BEAMnrc *Med. Phys.* **33** 3046–56
[39] Walters B, Rogers D and Kawrakow I 2009 *DOSXYZnrc users manual* (Ottawa, Canada: NRCC Report PIRS-794revB)
[40] Bazalova M and Graves E E 2011 The importance of tissue segmentation for dose calculations for kilovoltage radiation therapy *Med. Phys.* **38** 3039–49
[41] Briesmeister J 2000 *MCNP—A general Monte Carlo N-particle transport code, version 4C*LA-13709-M T R editor (Los Alamos: Los Alamos National Lab.)
[42] Pelowitz D MCNPX User's Manual, Version 2.7.0: ORNL/RSICC; 2011
[43] Chiavassa S, Lemosquet A, Aubineau-Laniece I, de Carlan L, Clairand I and Ferrer L *et al* 2005 Dosimetric comparison of Monte Carlo codes (EGS4, MCNP, MCNPX) considering

external and internal exposures of the Zubal phantom to electron and photon sources *Radiat. Prot. Dosimetry* **116** 631–5
[44] Baró J, Sempau J, Fernández-Varea J M and Salvat F 1995 PENELOPE: An algorithm for Monte Carlo simulation of the penetration and energy loss of electrons and positrons in matter *Nucl. Instrum. Meth. Phys. Res.* B **100** 31–46
[45] Sempau J, Sánchez-Reyes A, Salvat F, Oulad ben Tahar H and Jiang S B 2001 Fernández-Varea J M 2001 Monte Carlo simulation of electron beams from an accelerator head using PENELOPE *Phys. Med. Biol.* **46** 1163
[46] Almansa J, Salvat-Pujol F, Díaz-Londoño G, Carnicer A, Lallena A M and Salvat F 2016 PENGEOM—A general-purpose geometry package for Monte Carlo simulation of radiation transport in material systems defined by quadric surfaces *Comput. Phys. Commun.* **199** 102–13
[47] Brualla L, Salvat F and Palanco-Zamora R 2009 Efficient Monte Carlo simulation of multileaf collimators using geometry-related variance-reduction techniques *Phys. Med. Biol.* **54** 4131–49
[48] Sempau J, Wilderman S J and Bielajew A F 2000 DPM, a fast, accurate Monte Carlo code optimized for photon and electron radiotherapy treatment planning dose calculations *Phys. Med. Biol.* **45** 2263–91
[49] Rodriguez M, Sempau J and Brualla L 2013 PRIMO: A graphical environment for the Monte Carlo simulation of Varian and Elekta linacs *Strahlenther Onkol.* **189** 881–6
[50] Rodriguez M, Sempau J, Fogliata A, Cozzi L, Sauerwein W and Brualla L 2015 A geometrical model for the Monte Carlo simulation of the TrueBeam linac *Phys. Med. Biol.* **60** N219
[51] Agostinelli S, Allison J, Amako K, Apostolakis J, Araujo H and Arce P *et al* 2003 Geant4—a simulation toolkit *Nucl. Instrum. Meth. Phys. Res. A* **506** 250–303
[52] Allison J, Amako K, Apostolakis J, Araujo H, Arce P and Asai M *et al* 2006 Geant4 developments and applications *IEEE Trans. Nucl. Sci.* **53** 270–8
[53] Paganetti H, Jiang H, Parodi K, Slopsema R and Engelsman M 2008 Clinical implementation of full Monte Carlo dose calculation in proton beam therapy *Phys. Med. Biol.* **53** 4825–53
[54] Incerti S, Baldacchino G, Bernal M, Capra R, Champion C and Francis Z *et al* 2010 THE GEANT4-DNA PROJECT *Int. J. Model. Simul. Sci. Comput.* **01** 157–78
[55] Jan S, Santin G, Strul D, Staelens S, Assie K and Autret D *et al* 2004 GATE: A simulation toolkit for PET and SPECT *Phys. Med. Biol.* **49** 4543–61
[56] Sarrut D, Bardies M, Boussion N, Freud N, Jan S and Letang J M *et al* 2014 A review of the use and potential of the GATE Monte Carlo simulation code for radiation therapy and dosimetry applications *Med. Phys.* **41** 064301
[57] Chibani O and Ma C M 2007 On the discrepancies between Monte Carlo dose calculations and measurements for the 18 MV varian photon beam *Med. Phys.* **34** 1206–16
[58] Khan F M 2010 *The Physics of Radiation Therapy* (Lippincott Williams & Wilkins)
[59] Ahnesjo A and Aspradakis M M 1999 Dose calculations for external photon beams in radiotherapy *Phys. Med. Biol.* **44** R99–155
[60] Gibbons J P, Antolak J A, Followill D S, Huq M S, Klein E E and Lam K L *et al* 2014 Monitor unit calculations for external photon and electron beams: Report of the AAPM Therapy Physics Committee Task Group No. 71 *Med. Phys.* **41** 031501
[61] Ding G X 2002 Energy spectra, angular spread, fluence profiles and dose distributions of 6 and 18 MV photon beams: Results of Monte Carlo simulations for a varian 2100EX accelerator *Phys. Med. Biol.* **47** 1025–46

[62] IMRT CWG 2001 Intensity-modulated radiotherapy: Current status and issues of interest *Int. J. Radiat. Oncol. Biol. Phys.* **51** 880–914

[63] Mock U, Georg D, Bogner J, Auberger T and Potter R 2004 Treatment planning comparison of conventional, 3D conformal, and intensity-modulated photon (IMRT) and proton therapy for paranasal sinus carcinoma *Int. J. Radiat. Oncol. Biol. Phys.* **58** 147–54

[64] Sikora M, Dohm O and Alber M 2007 A virtual photon source model of an Elekta linear accelerator with integrated mini MLC for Monte Carlo based IMRT dose calculation *Phys. Med. Biol.* **52** 4449–63

[65] Sterpin E, Salvat F, Cravens R, Ruchala K, Olivera G H and Vynckier S 2008 Monte Carlo simulation of helical tomotherapy with PENELOPE *Phys. Med. Biol.* **53** 2161–80

[66] Lobo J and Popescu I A 2010 Two new DOSXYZnrc sources for 4D Monte Carlo simulations of continuously variable beam configurations, with applications to RapidArc, VMAT, TomoTherapy and CyberKnife *Phys. Med. Biol.* **55** 4431–43

[67] Bär W, Schwarz M, Alber M, Bos L J, Mijnheer B J and Rasch C *et al* A comparison of forward and inverse treatment planning for intensity-modulated radiotherapy of head and neck cancer *Radiother. Oncol.* **69** 251–8

[68] ICRU. Prescribing, Recording, and Reporting Intensity-Modulated Photon-Beam Therapy (IMRT). Report 83. 2010 1473-6691 Contract No.: 1.

[69] Granero D, Perez-Calatayud J and Ballester F 2007 Technical note: Dosimetric study of a new Co-60 source used in brachytherapy *Med. Phys.* **34** 3485–8

[70] Strohmaier S and Zwierzchowski G 2011 Comparison of (60)Co and (192)Ir sources in HDR brachytherapy *J. Contemp. Brachyther.* **3** 199–208

[71] Gill M R, Falzone N, Du Y and Vallis K A 2017 Targeted radionuclide therapy in combined-modality regimens *Lancet Oncol.* **18** e414–e23

[72] Zhang L, Chen H, Wang L, Liu T, Yeh J and Lu G *et al* 2010 Delivery of therapeutic radioisotopes using nanoparticle platforms: Potential benefit in systemic radiation therapy *Nanotechnol. Sci. Applic* **3** 159–70

[73] Hainfeld J F, Smilowitz H M, O'Connor M J, Dilmanian F A and Slatkin D N 2013 Gold nanoparticle imaging and radiotherapy of brain tumors in mice *Nanomedicine.* **8** 1601–9

[74] Felix M C, Fleckenstein J, Kirschner S, Hartmann L, Wenz F and Brockmann M A *et al* 2015 Image-guided radiotherapy using a modified industrial micro-CT for preclinical applications *PLoS One.* **10** e 0126246

[75] Wong J, Armour E, Kazanzides P, Iordachita I, Tryggestad E and Deng H *et al* 2008 A high resolution small animal radiation research platform (SARRP) with x-ray tomographic guidance capabilities *Int. J. Radiat. Oncol. Biol. Phys.* **71** 1591–9

[76] Newton J, Oldham M, Thomas A, Li Y, Adamovics J and Kirsch D G *et al* 2011 Commissioning a small-field biological irradiator using point, 2D, and 3D dosimetry techniques *Med. Phys.* **38** 6754–62

[77] Yahyanejad S, van Hoof S J, Theys J, Barbeau L M, Granton P V and Paesmans K *et al* 2015 An image guided small animal radiation therapy platform (SmART) to monitor glioblastoma progression and therapy response *Radiother. Oncol.* **116** 467–72

[78] Rodriguez M, Zhou H, Keall P and Graves E 2009 Commissioning of a novel microCT/RT system for small animal conformal radiotherapy *Phys. Med. Biol.* **54** 3727–40

[79] Hamdi M, Mimi M and Bentourkia M 2017 Comparison between X-rays spectra and their effective energies in small animal CT tomographic imaging and dosimetry *Australas. Phys. Eng. Sci. Med.* **40** 29–37

[80] Bazalova M, Nelson G, Noll J M and Graves E E 2014 Modality comparison for small animal radiotherapy: A simulation study *Med. Phys.* **41** 011710
[81] Hamdi M, Mimi M and Bentourkia M 2017 Impact of X-ray energy on absorbed dose assessed with Monte Carlo simulations in a mouse tumor and in nearest organs irradiated with kilovoltage X-ray beams *Cancer Radiother.* **21** 190–8
[82] Hui T E, Fisher D R, Kuhn J A, Williams L E, Nourigat C and Badger C C *et al* 1994 A mouse model for calculating cross-organ beta doses from yttrium-90-labeled immunoconjugates *Cancer* **73** 951–7
[83] Taschereau R and Chatziioannou A F 2007 Monte Carlo simulations of absorbed dose in a mouse phantom from 18-fluorine compounds *Med. Phys.* **34** 1026–36
[84] Mohammadi A and Kinase S 2011 Monte Carlo simulations of photon specific absorbed fractions in a mouse voxel phantom *Prog. Nucl. Sci. Technol.* **1** 126–9
[85] Kostou T, Papadimitroulas P, Loudos G and Kagadis G C 2016 A preclinical simulated dataset of S-values and investigation of the impact of rescaled organ masses using the MOBY phantom *Phys. Med. Biol.* **61** 2333–55
[86] Hindorf C, Ljungberg M and Strand S E 2004 Evaluation of parameters influencing S values in mouse dosimetry *J. Nucl. Med.* **45** 1960–5
[87] Larsson E, Ljungberg M, Strand S-E and Jönsson B-A 2011 Monte Carlo calculations of absorbed doses in tumours using a modified MOBY mouse phantom for pre-clinical dosimetry studies *Acta Oncol.* **50** 973–80
[88] Khoo A M, Cho S H, Reynoso F J, Aliru M, Aziz K and Bodd M *et al* 2017 Radiosensitization of prostate cancers *in vitro* and *in vivo* to erbium-filtered orthovoltage x-rays using actively targeted gold nanoparticles *Sci. Rep.* **7** 18044
[89] Cai Z, Kwon Y L and Reilly R M 2017 Monte Carlo N-Particle (MCNP) modeling of the cellular dosimetry of 64Cu: Comparison with MIRDcell S values and implications for studies of its cytotoxic effects *J. Nucl. Med.* **58** 339–45
[90] Paquette M, Vilera-Perez L-G, Beaudoin S, Ekindi-Ndongo N, Boudreaut P-L and Bonin M-A *et al* 2017 Targeting IL-5Rα with antibody-conjugates reveals a strategy for imaging and therapy for invasive bladder cancer *OncoImmunology.* **6** e 1331195
[91] Jalilian A and JO J 2017 The current status and future of theranostic Copper-64 radiopharmaceuticals **25** 1–10
[92] Humm J L 1987 A microdosimetric model of astatine-211 labeled antibodies for radioimmunotherapy *Int. J. Radiat. Oncol. Biol. Phys.* **13** 1767–73
[93] Andersson H, Elgqvist J, Horvath G, Hultborn R, Jacobsson L and Jensen H *et al* 2003 Astatine-211-labeled antibodies for treatment of disseminated ovarian cancer *Clin. Cancer Res.* **9** 3914s
[94] Pham Q T, Anne A, Bony M, Delage E, Donnarieix D and Dufaure A *et al* 2015 Coupling of Geant4-DNA physics models into the GATE Monte Carlo platform: Evaluation of radiation-induced damage for clinical and preclinical radiation therapy beams *Nucl. Instrum. Meth. Phys. Res. B* **353** 46–55
[95] Delage E, Pham Q T, Karamitros M, Payno H, Stepan V and Incerti S *et al* 2015 PDB4DNA: Implementation of DNA geometry from the Protein Data Bank (PDB) description for Geant4-DNA Monte-Carlo simulations *Comput. Phys. Commun.* **192** 282–8

IOP Publishing

Computational Anatomical Animal Models
Methodological developments and research applications
Habib Zaidi

Chapter 14

Summary and future outlook

Habib Zaidi

14.1 Summary

This is an exciting time for preclinical multimodality molecular imaging-based biomedical research [1]. The combination of realistic computational animal models with Monte Carlo simulations has become one of the most popular tools in a variety of medical physics research fields subsequent to the availability of advanced modeling mechanisms and powerful computing resources [2]. A large number of computational animal models have been developed over the last two decades dedicated mostly to preclinical imaging physics and radiation dosimetry studies [3, 4]. This logical evolution has been driven by a number of fundamental and critical developments described in the 13 chapters of this volume reporting on the conceptual design of computational animals models (see chapters 1 and 2), a detailed description of models (chapters 3–8) and simulation tools (chapter 9) reported in the literature as well as the different applications of these models in biomedical research (chapters 10–13).

The utilization of these models is expected to grow as advanced and fast simulation tools are developed and adopted as validated approaches in various biomedical research applications involving the use of laboratory animals. Since 1990, more than 31 212 PUBMED entries have appeared in the literature (about half of them published during the last 5 years alone) when searching for ('ANIMAL' OR 'RODENT' OR 'MOUSE' OR 'RAT') AND ('MODEL' OR 'PHANTOM') AND ('IMAGING' OR 'DOSIMETRY' (figure 14.1). It should be noted that this search does not indicate that there are 31 212 papers that deal with animal computational models, but only that the exponential increase of the occurrence of these words is an indicator of the rapidly growing field of computational modeling in the preclinical field (an increase of ~2340% from 1990 to 2017).

This volume is dedicated to the challenges associated with the conceptual design of realistic computational animal models in connection with their use as input for simulation tools. Their research applications are also addressed in great detail. This

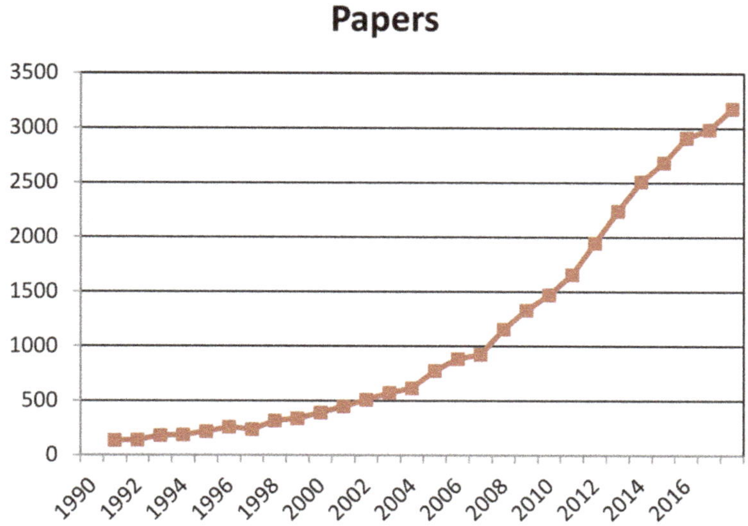

Figure 14.1. The increasing number of annual peer-reviewed publications reporting on the use of small-animal models or phantoms in preclinical research demonstrates the growing interest in these tools. This graph is based on a PubMed query with using the following mesh terms: ('ANIMAL' OR 'RODENT' OR 'MOUSE' OR 'RAT') AND ('MODEL' OR 'PHANTOM') AND ('IMAGING' OR 'DOSIMETRY') for each year from 1990 to 2017.

volume can be regarded as a preliminary response to the remarkable challenges in the field and may be viewed as an early album containing several of the many and varied snapshots of this rapidly growing area.

14.2 Future outlook

It is gratifying to see in perspective the progress that computational object modeling has made, from simple stylized models of reference animals, through voxel-based models and most recently towards truly subject-specific models based on actual animal images and advanced boundary representation techniques.

It is anticipated that over the next 5–10 years, the majority of Monte Carlo radiation transport software packages will be employing dynamic 4D boundary representation animal models to respond to the growing call for more realistic modeling. Therefore, it is crucial that preclinical imaging and radiation dosimetry communities establish guidelines defining the requirements for simulation studies to ascertain applications where stylized models are appropriate and those applications necessitating more refined voxel- or boundary representation-based computational models to produce optimum results. These are challenges that should be addressed soon to enable the widespread use of these tools in biomedical research.

This is also an exciting time for computational modeling of the anatomy and physiological functions of laboratory animals used extensively in biomedical research. During the last 20 years, the number of published papers on this topic has been rising gradually, which motivated us to pull together this volume dedicated to computational animal models as a snapshot of this dynamically shifting field. The growth of computational animal models has been very fast and stimulating, and there is every reason to predict that the field will move forward more quickly in the foreseeable future, taking advantage of the advent of powerful computing resources and modeling methodologies and the infinite imagination of scientists. In spite of the outstanding accomplishments summarized in this volume, there is still scope for additional research. There is no shortage of challenges and opportunities for computational animal models at the present time. We hope that in this limited space we were able to offer you a flavor of up-to-date advances in computational animal models and their impending applications in preclinical research. We found the compilation of this volume to be a rewarding and educational experience and hope that the reader is left with the same experience.

14.3 Acknowledgements

This work was supported by the Swiss National Science Foundation under grant SNSF 31003A-149957 and the Swiss Cancer Research Foundation under Grant KFS-3855-02-2016.

References

[1] Zaidi H (ed) 2014 *Molecular Imaging of Small Animals: Instrumentation and Applications* (New York: Springer)

[2] Zaidi H 1999 Relevance of accurate Monte Carlo modeling in nuclear medical imaging *Med. Phys.* **26** 574–608

[3] Xie T and Zaidi H 2016 Development of computational small animal models and their applications in preclinical imaging and therapy research *Med. Phys.* **43** 111–31

[4] Zaidi H and Xu X G 2007 Computational anthropomorphic models of the human anatomy: The path to realistic Monte Carlo modeling in medical imaging *Annu. Rev. Biomed. Eng.* **9** 471–500

Lightning Source UK Ltd.
Milton Keynes UK
UKHW050620060821
388415UK00003B/72

9 780750 31